Make: Trigonometry

Build Your Way from Triangles to Analytic Geometry

By Joan Horvath and Rich Cameron

Dedication

To the open source 3D printing community, for making consumer 3D printing possible.

Make: Trigonometry

by Joan Horvath and Rich Cameron

ISBN: 978-1-68045-798-8

September 2023: First Edition

See **www.oreilly.com/catalog/errata.csp?isbn=9781680457988** for release details.

Make: Books

President Dale Dougherty

Creative Director (cover) Juliann Brown

Editor Kevin Toyama

Art Director/Photographer/Designer Rich Cameron

Technical Reviewer Dr. Niles Ritter

Proofreader Doug Adrianson

Make Community, LLC

150 Todd Road, Suite 100

Santa Rosa, California 95407

www.make.co

Table of Contents

Preface

This book covers two closely related topics: trigonometry and analytic geometry. You will get more out of this book if you read a geometry book first, whether it be our *Make: Geometry* or another book that covers triangles, circles, polyhedra, cross-sectional slices, the Pythagorean Theorem, and similar topics. This book, in turn, covers some of the topics that are helpful background for our *Make: Calculus*.

Who This Book Is For

This is not a standard algebra II, trigonometry, or precalculus textbook. Rather, we have pulled out concepts that we think are more intuitive if taught first in a hands-on, applied way. If you are taking what in the United States would be a high school course called "Algebra II and Trigonometry," you should find this book covering some (but not all) of the same ground.

Analytic geometry is all about tying together geometric curves and the equations (and graphs) that represent them. There are two ways to go about this: either geometry-first or equations-first. You can probably guess that we are going about this geometry-first. Analytic geometry is typically seen as a run-up to calculus, and so this material might help you align well with our *Make: Calculus* book.

Teaching and Learning with This Book

It is our belief that teaching with 3D objects first and then returning to abstractions gives a better foundation and more intuition for higher-level math. Thus we do not claim to cover all of any grade level's required math in any jurisdiction. We cover key concepts that lend themselves to our approach. We list topics covered in each chapter in an Appendix for ready reference.

Our approach broadly follows ideas that have been called, among other things, constructivism (in the educator's sense of the word), problem-based learning, and active learning. We have been influenced by Paul Lockhart's work, particularly his books, *The Mathematician's Lament* (Belknap, 2009) and *Measurement* (Belknap, 2012). We also appreciate the work of Jo Boaler, particularly her book *Mathematical Mindsets* (Jossey-Bass, 2016).

The 3D printable models in the book were written in **OpenSCAD** (https://www.openscad.org/), a free, open-source CAD program. If you do not have a 3D printer readily available, visualizing and manipulating the models in the OpenSCAD environment will provide some of the experience.

The models visualize scaffolded learning concepts that might not be in the "normal" sequence, but are suited to hands-on learning. Ideally, students will get into OpenSCAD and mess with the models a bit, as we describe when we introduce them. Most of them are not intended as one-off visualizations, but rather as starting points to be modified and explored.

What You Will Need

This book assumes some prior knowledge of algebra and geometry. In Chapter 1, we briefly review concepts that we assume you have seen by this point. We will make sure you at least know what these concepts are called, so you can look up the details if any of it is new.

OpenSCAD runs on laptop or desktop computers running Windows, macOS, or Linux. As of this writing, OpenSCAD does not run on Chromebooks or tablets. (There are versions that do, but they do not support functionality we need for our models, and some also require a paid license.) Creating models in OpenSCAD from scratch requires learning computer programming skills. However, here you will be working with models we created for you, which just involves changing a few numbers or an equation in the model. We give enough pointers that someone without that background should be able to manage those basics.

We encourage you to print the models on any consumer-level 3D printer that uses filament. (Printers that use liquid resin might require that you tweak the models a bit.) As we discuss in Chapter 2, decent printers can be found under $500, and many community libraries have printers available. Some of the models can also be printed on paper and assembled. However, if that simply is not an option, you can open the models in OpenSCAD and manipulate them in 3D to get some insight.

In some cases, though, we have several OpenSCAD models that are used together. Unfortunately, OpenSCAD does not have a good way to support that on the screen without printing out physical models. In those cases, you still should be able to read this book and understand the concepts from our photographs as a last resort.

Other than a 3D printer, we have tried to stick to supplies that can be purchased cheaply or cobbled together. Where possible, we also provide examples you can create with common household items, such as paper, pipe cleaners, bits of wire, and so on, and slightly more exotic ones, like a protractor and straw. There is a list of downloadable models and physical materials needed at the start of each chapter.

The one project that requires both a 3D printer and some specialized materials is the Arduino robot arm project in Chapter 12. This project assumes some knowledge of open-source electronics like Arduino microcontrollers and hobby servos. If you choose to just read that chapter and understand how we programmed the arm's motion, you should get a lot of the math benefit, although you will not be able to play with it in quite the same way.

As a final note, the math in this book was developed in MathML, which is designed to create screen-reader-accessible math in an epub file. The **Maker Shed** (http://www.makershed.com) will carry a package of both pdf and epub versions. The open source **Thorium** (https://www.edrlab.org/software/thorium-reader/) epub reader is one free, compatible option.

After downloading Thorium, open this book in it. After you have the book open, turn on MathJax functionality, which in turn reads MathML. In the current version of Thorium (2.2.0) check the box to enable MathJax at `Settings` > `Display` > `MathJax`. Otherwise Thorium will try to read the markup version aloud. These settings are only available after you have opened at least one book in Thorium, though. If you are using a separate screen reader program to interface with Thorium, you might need to add a plugin for it to be able to read MathML correctly.

Acknowledgments

To write a book like this requires building on mathematical insights that have been evolving for millennia. We wish that we could have met some of the mathematicians who lived 2,000 years ago in ancient Greece, or some of those in Renaissance Europe. We like to think that many of them might have been intrigued to see ideas of theirs made concrete with a 3D printer.

We are not able to go back in time to chat with Archimedes or Galileo, but we can thank the many people present today who helped us. It is never possible to capture everyone who contributed, so we apologize in advance for anyone we miss here. Your contributions were still critical.

Tech reviewer Dr. Niles Ritter brought a mathematician's mindset to his review of this book. We particularly appreciated his historical notes that enriched the final project significantly, and his suggestion that we look into the monotile research we incorporated in Chapter 12.

We shamelessly pestered other techie friends as informal sounding boards as well. Chief among these was Joan's husband Stephen Unwin, who was very helpful in helping us think about some of the more physics-oriented questions.

We thank all the staff at Make: Community LLC, who supported the many pieces that go into producing a book like this one. In particular we appreciate editors Michelle Lowman, who helped us get this project started, and later Kevin Toyama, who saw it through to fruition. Publisher Dale Dougherty graciously supported this endeavor and our earlier ones. Old friend Doug Adrianson was the eagle-eyed proofreader. Any remaining errors, of course, remain ours.

We also want to note that developing some of the material for this book was supported in part by grant number 90RE5024, from the U.S. Administration for Community Living, Department of Health and Human Services, Washington, D.C. 20201. We also are grateful for the insights of Yue-Ting Siu into creating tactile models and the Smith-Kettlewell Eye Institute for their support.

Finally, Christian Lawson-Perfect allowed us to point to his Github repository in Chapter 12 so you can use his marvelous set of printable and laser-cuttable mathematical tiles.

Most photographs and images were created by the authors. Images from other sources are credited in their captions.

About the Authors

Joan is an MIT alumna and recovering rocket scientist. She worked on spacecraft to several planets in the first part of her career. Rich (known online as "Whosawhatsis") is an experienced open source developer who has been a key member of the RepRap 3D printer development community for many years. His designs include the original spring/lever extruder mechanism used on many 3D printers, the RepRap Wallace, and the Deezmaker Bukito portable 3D printer.

Together, we run a Pasadena-based consulting and training firm, **Nonscriptum LLC** (https://www.nonscriptum.com) which focuses on teaching educators and scientists how to use maker technologies like 3D printing and open-source electronics. This is our tenth book together, and we have also developed courses on additive manufacturing for LinkedIn Learning.

As we taught people the nuts and bolts of using 3D printers in a classroom, we found many educators started with existing textbooks and tried to bolt on "making something." The more we explored it and combined Joan's traditional education with Rich's learn-by-making mindset, the more we were convinced it made sense to take a subject and imagine how to start over with hands-on learning in mind. We would love to hear about your experiences teaching or learning with this book. You can reach us through our **website contact page** (https://www.nonscriptum.com/contact).

1 Trigonometry and Analytic Geometry

If you ask someone what trigonometry is, they will probably mutter something about angles, sides, and hypotenuses, and perhaps some strange acronyms they memorized long ago. This is pretty ignominious for a discipline that enabled navigation across oceans, and, eventually, across interplanetary space. Hopefully, by the time we are through this book, you will be able to give a more compelling definition. Let's start our book by getting a little better idea of what trigonometry and analytic geometry actually are, and talking about why we have combined them in this book. Then we will walk through what to expect as you travel with us, and what we assume you already know.

Trigonometry

Trigonometry was developed for very practical navigation and surveying applications. When we ride on a straight railway track, or have our phones use satellite data to get us somewhere, we are benefiting from trigonometry (or "trig" to its friends). We will start our study of trig looking at the properties of just one triangle, and seeing the relationships among the angles and sides of triangles, particularly triangles with one 90° angle.

Until the cheap-calculator era, people who were doing practical things needing trig would have to carry around books consisting of tables of numbers ("trig tables") in which they would look up values they needed. We do not have to do that anymore. But, as we will see, a fair amount of interesting math was developed along the way to make calculation and using those tables less painful. It is thought that the first person to generate forerunners of trig tables and use them to solve problems was the Greek astronomer Hipparchus, who lived about 2100 years ago. The ancient Greeks computed trigonometric quantities differently than we will learn in Chapter 3, but the premise was equivalent.

Much later, people discovered interesting properties of the curves generated by plotting the value of trig functions as the angles of a triangle changed. These curves, which look like waves, have properties that have given us models of many physical phenomena, from water waves to swinging pendulums to propagation of light. We will learn about these waves and their applications in Chapters 5 through 8. All of it, though, can be derived by starting with a triangle, as we will see.

Analytic Geometry

We are so used to seeing graphs in mathematics books (and textbooks in other subjects) that it might seem that the idea of graphing a curve must be ancient. However, the common two-dimensional graph with axes at right angles was developed not long after the *Mayflower* sailed to America. The French philosopher René Descartes at some point during the 1620s realized that one could create a gridwork we would now call a coordinate system, and use it to plot curves. The common x-y Cartesian coordinate system now bears his name.

The creation of coordinate systems tied geometrical abstract ideas like triangles and circles to equations, allowing for calculations and analysis that would not have been possible without them. Isaac Newton would have had a hard time developing calculus about two decades later if Descartes had not invented coordinate systems and means of using them when he did.

Analytic geometry broadly involves thinking about geometric constructs as they are translated into graphs in a coordinate system. In Chapter 4, we learn about coordinate systems, and how to change from one to another, since it is sometimes more convenient to use something other than a Cartesian grid to work on problems.

Most of our analytic geometry section of the book (primarily Chapters 9, 10, and 11) shows where equations for circles, ellipses, parabolas, and hyperbolas come from. These are all shapes that result from cutting a cone at various angles, as you may have learned in a geometry class, and we will reprise that here. Some very powerful math falls out of these explorations, as well as explanations for how some kinds of telescopes and solar collectors work.

Structure of This Book

Each chapter starts with a list of what is needed to do the activities in that chapter, including any 3D printable files. Then, we introduce the math content, with a physical model or experiment if possible. Finally, we summarize the chapter and list terms that the reader can look up to learn more. We know that self-learners often get stuck because they do not know what something is called, so we make sure that you know the lingo.

We start off in Chapter 2 with an introduction to the OpenSCAD computer-aided design (CAD) program. This is a free, open-source program that we used to design all the models for this book. Chapter 2 tells you how to install and use OpenSCAD, as well as giving directions to download the 3D printable model files. The models, too, are freely available, as long as they are attributed to this book's repository.

Then, Chapter 3 begins our exploration of trigonometry in the context of a single triangle. We learn the basic trigonometric functions (sine, cosine, and tangent) in a right triangle. Next, we see how to use properties of triangles to measure objects impractical to measure directly. After we get some more background in the intervening chapters, we will reprise these ideas in Chapter 7 and learn more about navigation. There we show you how to make a very simple inclinometer and to use it to measure your latitude with the North Star.

In Chapter 4, we have a side excursion into coordinate systems, which sets us up both for an expansion of trigonometry beyond one kind of triangle and for our analytic geometry explorations. Chapters 2, 3, and 4 have some commonality with our *Make: Geometry* book, and if you have worked with that book these chapters will mostly be a refresher for you.

It might seem that a circle with a radius of 1 is a boring construct to get excited about, but it turns out that this "unit circle" is the key to moving trigonometry from the angles within one triangle to continuous curves of sine, cosine, and the rest of the functions. Chapter 5, with the help of some 3D prints and a bit of Play-Doh, will help us walk through this magic.

In Chapter 6, we learn how studying trigonometry contributed, in a roundabout way, to logarithms and slide rules. Slide rules are simple and useful for getting all kinds of insight, and we create a simple 3D printable minimalist slide rule, which can also be printed on paper instead if a 3D printer is not available. Chapter 8 explores the applications of sine and cosine curves to problems like predicting how light will bend when it hits the surface of a body of water.

Our explorations of analytic geometry, beyond being introduced to coordinate systems in Chapter 4, are focused in Chapters 9, 10, and 11. We start with geometrical definitions of the circle, ellipse, parabola, and hyperbola, and then use those definitions to derive equations for them. Then, once we have those explanations, we see where they take us, both mathematically and with applications. The early parts of each chapter are reprises of their derivations in *Make: Geometry*, but then we take them further, and in different directions.

Finally, we have two categories of projects to explore in Chapter 12. These projects are intended to be open-ended. The first project explores tiling, finding shapes that (like bathroom tiles) cover a flat surface completely. We report on some very new work on a tile shape that can completely cover a surface without ever repeating patterns, and point you to a free source for ways to make this tiling with a 3D printer, laser cutter, or on paper.

The other project shows you how to construct and program a small toy robot arm. We use an Arduino microprocessor and hobby servos to drive 3D printed arm segments. Information on how to program an Arduino processor is widely available, and we point you to it. But, if you prefer not to take a side stroll through electronics, you can just follow along and learn the trig involved in telling a simple robot where to put its arm.

This last chapter ends with a few resources you might find useful as you go forward in your math career, or to explore topics in this book in more depth. We have also included an Appendix which lists the topics included in each chapter. If you are teaching from this book, you should be able to translate from these to your jurisdiction's standards or other requirements.

Math Concepts to Know

By and large, we will define mathematical terms as we go. However, we expect that readers have seen basic algebra and some geometry before wading into these topics. (Ideally you will have read our *Make: Geometry* book, or its equivalent.) By "basic" we mean that you should know

- Algebra fundamentals, including understanding positive and negative numbers, fractions, and being able to solve for x in an equation like $3x - 5 = 4$.
- The different conventions for math symbols in algebra, computer code, and arithmetic. For example, algebra uses an implied "times sign" by writing letters or numbers next to each other. However, computer code uses a " * " to show multiplication. We also use this symbol in some of the equations in this book, when we want to show multiplication explicitly.
  - Arithmetic: $y = a \times x$, $y = 2 \times x$, $(x + 3) \times (x + 3)$
  - Algebra: $y = ax$, $y = 2x$, $(x + 3)(x + 3)$
  - OpenSCAD programming language: `y = a * x; y = 2 * x; (x + 3) * (x + 3)`
- What it means to raise something to a power, including a negative power. (Negative powers are in the denominator of a fraction.) Look up the "power rule for exponents" for more.
  - Examples: $x^3 = x$ times x times x, and $x^{-3} = \dfrac{1}{x^3}$
- How to manipulate fractions that include algebra. (For example, simplifying algebraic expressions in fractions using least common denominator principles and similar strategies). Search for "adding algebraic fractions" for details.
  - Example: $\dfrac{1}{x+3} + \dfrac{1}{2} = \dfrac{2}{2(x+3)} + \dfrac{x+3}{2(x+3)} = \dfrac{2+x+3}{2(x+3)} = \dfrac{x+5}{2(x+3)}$
- How to multiply two polynomials (expressions with variables) together. If you are shaky about this, there is a sidebar in Chapter 10 with details about the FOIL method for doing this.
  - Example: $(x + 1)(2x + 3) = 2x^2 + 5x + 3$
  - How to reverse this to factor a polynomial. That is, start with $2x^2 + 5x + 3$ and find its factors $(x + 1)(2x + 3)$.
- What square roots, cube roots, and other roots are; how to think of roots as raising a variable to a fractional power; and how to interpret the radical sign. If you are a little unclear here, we will go into this more in Chapter 6.
  - Example: square root: $x^{1/2} = \sqrt{2}$. cube root: $x^{1/3} = \sqrt[3]{x}$.
- That the Greek letter pi (π) is the ratio between the circumference of a circle and its diameter, and that it is an irrational number. The Wikipedia article "Pi", is a good comprehensive source, covering the next three points as well.

- What an angle is, how to measure an angle (degrees and radians), and the number of degrees in a circle (360°) and the sum of the vertices of a triangle (180°).
- That 2π radians = 360°, and why that is true.
- That an irrational number is a number that cannot be expressed as a ratio of two integers.
- How to compute areas of squares, circles, and similar 2D shapes, and volumes of basic 3D shapes (sphere, cylinder, cube). Search on "area formulas" and "volume formulas" to find specific cases.

Should you hit other terms that are unfamiliar later in the book, the **Khan Academy** (https://www.khanacademy.org) is a well-organized place to look for background on almost any mathematics subject. **Wikipedia** (https://wikipedia.org) is usually good too, but can sometimes be a little dense in its presentation. Random online searches, as always, might be the easy way to go, but the results can be uneven.

Chapter Key Points

In this chapter, we briefly introduced trigonometry and analytic geometry. We walked through the topics in the rest of the book. Finally, we summarized the mathematics you should already know to use this book effectively. We also gave suggestions on where to brush up your skills if need be.

2

3D Printed Models

Resources used in this chapter

The OpenSCAD program (download from **OpenSCAD.org** (https://openscad.org/))

The following 3D printable models are used in this chapter. Download instructions are in the text.

- `axes.scad`
  - This model creates various coordinate planes, and a set of coordinate axes.

Throughout the book, we illustrate concepts with a variety of hands-on projects, including a lot of 3D printed models. We assume you are comfortable with the basics of 3D printing. If not, there are many resources out there. We have listed some of ours in the References if you need some help getting started. At the end of this chapter, we briefly review the overall 3D printing process, with a focus on what you might need to tweak to print some of the models more efficiently.

The models have been created in an open source computer aided design (CAD) program called OpenSCAD. OpenSCAD allows you to create models by writing computer code in a language similar to C or Java, with some exceptions. The models have been designed so that you can just use the model as-is for a basic case. To explore more, you can change a parameter value or two, or perhaps modify an equation. Each chapter explains what to do to alter its models, based on a general familiarity with OpenSCAD and 3D printing from this chapter.

If you want a deeper dive into using OpenSCAD for mathematical modeling, you might check out the first three chapters of our 2021 book *Make: Geometry*. We go into considerable depth there on both the mechanics of OpenSCAD and its inherent applicability for teaching geometrical concepts.

In this chapter, we talk about OpenSCAD and its workflow first, and then discuss how it fits into a 3D printing workflow. We also give alternatives for using OpenSCAD as a simulator if you do not have access to a printer.

OpenSCAD

OpenSCAD is an open source computer aided design (CAD) program that is available for **free download** (https://openscad.org). It is a constructive-geometry program, designed for 3D printing. Creating a model starts with either 2D primitive shapes (like circles) or 3D ones (like spheres) to build up more complex shapes. Models are built in a language that is similar to programming languages in the C/Java/Python family. The models in this book assume you are using OpenSCAD version 2021.01 or later.

To use OpenSCAD, first **download it** (https://openscad.org) and follow any on-screen instructions to install it. The manual (under the "documentation" tab on the website) has a very good and inclusive description of how to use the program and its associated modeling language. We summarize the key points here, but we suggest you refer to the manual as well. We show menu items and OpenSCAD code `in a different font`, here and elsewhere in the book.

OpenSCAD has four basic panes (Figure 2-1).

- Type (or paste in) the text of your model in the *Editor*.
- The *Console* will tell you if you have issues when your print is rendered and is also where any messages that you write into your models will appear.
- The *Customizer* allows easier changes to code that is designed for it.
- Finally, there is a *display window* that displays the 3D model when you preview or render it.

If you do not see one or more of these windows, open the `View` menu and see if they have been hidden. A checkmark will appear next to "hide (name of window)" if so. Uncheck to show the window, or check the menu item to hide it. You can resize various panes within the window by dragging the borders between them.

Everything in OpenSCAD has to be input as text; there is no edit capability in the display window. You can pan, rotate, and zoom the view to look at different aspects of your model, but not resize or otherwise alter it.

FIGURE 2-1

OpenSCAD basic interface

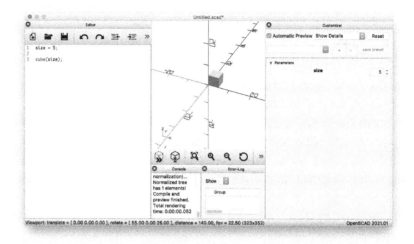

OpenSCAD Workflow

There are menu items listed across the top of the OpenSCAD screen. Alternatively, you can use various buttons and shortcuts described in the manual. We use the convention `Menu > Item` to say, "Click `Menu`, then click the `Item` shown in the pulldown menu that appears." Here are the basic functions you need to create or modify a model, and export it to 3D print.

OpenSCAD creates an STL file, which is then an input into either an open source or proprietary "slicing" program, depending on your 3D printer. (STL stands either for "stereolithography" or "standard tessellation language", depending on whom you ask.) See the section in this chapter about 3D printing for more information about these files.

OpenSCAD models are saved in `.scad` files. Look at the `.scad` files if you want to alter the model. Many different models in this book may have been made from one `.scad` file, by changing some parameters or a few lines of the model. At the start of each chapter, we note which figures were created by which .scad and/or STL files.

Since OpenSCAD does crash occasionally, it is prudent to use `File > Save` before trying any of the other functions here to save an OpenSCAD-editable (`.scad`) file. OpenSCAD does not autosave, and it is easy to lose your entire model if the program crashes or hangs during preview or render.

- To preview (display a fast, imperfect version) use `Design > Preview`.
- To render (creates an accurate model, but usually takes longer than the preview; required for exporting an STL) use `Design > Render`.
- To export an STL file after rendering, use `File > Export > Export as STL`.
- If one of the panes is not visible, go to the `Window` menu and make certain that the respective pane is not hidden.
- To display and then export an STL file by copying and pasting from Github (also see the section "Downloading the Models: Github" later in this chapter):
  - `File > New`, then paste in the text from the Github model
  - `File > Save As...`
  - `Design > Preview`
  - Assuming you like what you see, `Design > Render`
  - `File > Export > Export as STL`

Finally, OpenSCAD has examples (not all of which are 3D printable) that you can find in `File > Examples`.

If you dig into the models in this book, you will probably be surprised at how simple many of them are. OpenSCAD is a very powerful language, once you get past its quirks, and you can create quite complex models with surprisingly little programming.

We used OpenSCAD version 2021.01 on Macs running OS X. Earlier versions of OpenSCAD may not support all the capabilities we use in this book, but models should run on Windows or Linux without issues. On a Mac with a recent operating system, you will need to go into System `Preferences > Security and Privacy` and approve installing OpenSCAD.

Navigating on the Screen

Once you have previewed or rendered a model, you can "fly around" it to view it in 3D. Click on the display window, and then use your mouse or trackpad as follows.

- To move around in general, hold down your left mouse button (Windows) or single mouse button (Mac) and move around as you desire.
- To shift the field of view, right-click (Windows) or Control-click (Mac) and drag.
- To zoom in or out, use your scroll wheel, or the magnifying glass buttons along the bottom of the preview pane.

Comments

Comment lines in OpenSCAD either start with `//` or are enclosed (if multiple lines) by `/*` at the start of a comment and `*/` at the end. To "comment out" (ignore) some code, either put `//` at the start of a line, or enclose several lines like this:

```
/* (your code) */
```

Conversely, remove such annotations to make code "live." OpenSCAD changes the color of comments and text, so you can see what code is live that way. (The respective colors depend on the color scheme you chose, but most color schemes use a green or gray tone for comments.) In the next section of this chapter, we go into detail about how to change the models.

Idiosyncrasies of OpenSCAD

There are a few significant differences from other programming languages. First, OpenSCAD does not have true "variables." It has something much closer to what other languages would call "constants." For example, if we say

```
xyz = 2;
```

at the top of a program, and later try

```
xyz = xyz + 5;
```

OpenSCAD will give unexpected results at runtime.

However, if you do this:

```
xyz = 2;
cube(xyz);
xyz = 5;
```

the last line will supersede the first, and you will get a 5 mm cube (and a warning in the console about the value being overwritten). OpenSCAD will use the last value it sees (in this case, making `xyz` equal to 5) for the entirety of the program. In a lot of our models, you will see a long string of values for the same constant, commented to show which figure those values correspond to. You might think of these as variables in the algebraic sense (so the variable has one value wherever you see it), rather than the programming sense (where a variable represents a piece of memory that can be written and rewritten as the program progresses).

This lack of variables that can be changed also makes it difficult to calculate values based on previously calculated ones. In other languages, it would be common, for example, to create a variable (which you might call "sum"), then loop through the values in a list, adding each value to the sum variable. When the loop finished, the variable would tell you the sum of all the values in the list.

In OpenSCAD, there is no way to overwrite the "sum" variable with a value based on its previous value. We could create a separate variable for each step in the process, but that would require us to write one (nearly identical) line of code creating a unique variable for each step in the process. Arrays are a good way to store lists of values like this, but to create an array in OpenSCAD (often referred to as vectors, since their most common use in OpenSCAD is to store a set of `[x, y]` or `[x, y, z]` values), we have to set all of the values at once, without being able to look at the previous values as we do it.

The solution, which many of our models use in one way or another, is to create recursive functions. These functions call themselves, and calculate a value based on the previous output of the same function, with different parameters. In that way, you can get the sum of an array by asking for the sum of the first value plus the sum of all values from the second onward, using the same function. When you ask for the sum of all values from the second onward, it adds the second value to the sum of all values from the third onward, and the process repeats.

Eventually, you will get to the last value in the array (which is the sum of all values from that one onward), and the recursion ends. OpenSCAD is careful to only allow this type of recursion if you are passing the function some kind of counter argument that increments each step and will eventually end the recursion, so that you do not get an infinite loop.

We also often use this recursion to build arrays. When this happens, the final recursion usually returns a single-element array (or an empty one), while the function otherwise returns an array that it creates by concatenating a calculated value with an array produced by the function calling itself. This way, each recursion returns an array that is one element longer than the one inside it.

Like variables, OpenSCAD's functions are functions in the mathematical sense, rather than the programming sense. Each function is a single expression, which can contain mathematical operators and other functions that return a value, but no looping. Branching within these functions is possible using the ternary operator, which works like an "if-else" statement inside an expression.

This looks something like `(a) ? b : c`. If `a` is an expression or value that evaluates as true (like a true comparison, or any non-zero number), the expression returns the value of `b`. If `a` is false, the expression instead returns `c`. Recursive functions in OpenSCAD always need to use the ternary operator so that they branch in a way that allows the recursion to end.

The Models

We have designed the models so that there are parameters that can be changed if you want to change the details of the models. Some of the models are used in multiple chapters. In those cases, we have comments showing the adaptations for different models and the alternative parameters commented out. Just change what is commented out (as described in the previous section) to create a different file. See the **OpenSCAD manual** (https://openscad.org/documentation) for details.

Each chapter notes how to change models when we first introduce them. If the model is set up for the Customizer, it will say so; otherwise assume that you are going to be changing a number, or perhaps changing which lines are commented out.

The Customizer makes it a bit easier for the user to change models with minimal coding, but it will limit you to a predefined set of parameters instead of letting you (in principle) change anything you want in a model. Models built to use the Customizer can still be edited without using it, but you may have to make all subsequent changes by editing the code (without the Customizer.)

The models for this book are supplied as directly 3D printable (STL) models, and as alterable CAD (.scad) files. First, we show a very simple model and change a parameter (the process for altering most of the models in the book). Then we show changing a different model by using the Customizer. You cannot edit an STL file in OpenSCAD.

Example 1: Changing a Parameter

First, let's create a minimal OpenSCAD model that creates a sphere sitting on top of a rectangular box. To make a box (a rectangular solid if you prefer) we use the OpenSCAD `cube()` module. By default, the `cube()` module creates a cube one millimeter on a side, with one corner at the origin. We can modify these dimensions (making it no longer a true cube, but still a rectangular solid) by putting a vector inside the parentheses to set its three dimensions.

The `sphere()` module creates a sphere centered on the origin, and we can put a number inside the parentheses to set its radius. To move the sphere to be on top of the box, we have to use the `translate()` module, which also takes a vector to specify how far to move an object in the x, y, and z directions. Here is the OpenSCAD to create and stack these shapes.

```
box_height = 10;
box_width = 20;
box_depth = 25;
sphere_radius = 5;
cube([box_width, box_depth, box_height]);
translate([box_width / 2, box_depth / 2, box_height + sphere_radius])
  sphere(sphere_radius);
```

Figure 2-2 shows the OpenSCAD rendering of this model. We could make a 10 by 20 by 25 mm box by using `cube([10, 20, 25])`. However, this means that if we also wanted to have a relationship between the position of the two boxes, we would have to remember to change the parameter in two places.

FIGURE 2-2

A sphere on top of a box

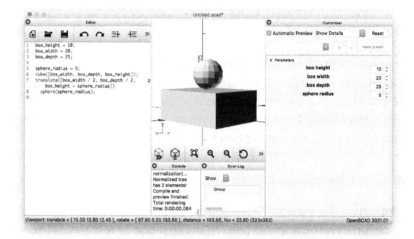

Having a parameter like this allows for keeping track of relationships among the variables. To make the bottom box taller and raise the sphere up to rest on top of it, we would just change the value of `box_height` to the new value, and both those things would happen automatically (Figure 2-3). This is a very simplistic example (and not included in the repository). Create a few examples like this to get comfortable with OpenSCAD's conventions before trying to alter the models later in the book.

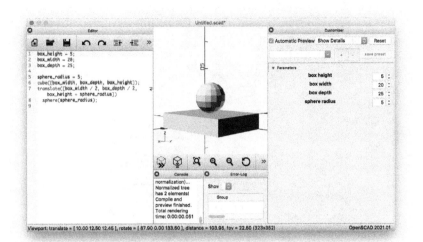

FIGURE 2-3

After changing box_height from 10 to 5 mm

Example 2: Changing a Model with the Customizer

Changing a model parameter or two is not difficult, as we have just seen. You just have to edit, save, and re-render the model, and then export a new STL file for 3D printing. However, some of the models, like Chapter 4's `axes.scad` model of coordinate systems, require the user to change several parameters for each case. Changing many parameters and commenting out some lines and uncommenting others gets unwieldy.

Fortunately, OpenSCAD has a capability called the *Customizer,* which allows the creator of a model to predefine items in pulldown menus. This allows for simpler, more-controlled editing by a user, but at the cost of flexibility. If a parameter does not appear in a menu item, it can only be changed by editing the model directly, not by using the Customizer. Figure 2-4 is an OpenSCAD screenshot of `axes.scad` showing the Customizer

We will note in comments in the model if a model assumes you are using the Customizer. If we have not noted that it does, behavior may be unpredictable if you try to change variables there. Variables will show up in the Customizer if they are set to a constant (not to an equation) and if they are ahead of the first curly brackets in the model. In some cases, we have added an empty set of curly brackets early in the model to suppress visibility of variables it would be a bad idea to change.

FIGURE 2-4

*Customizer
parameters for
axes.scad*

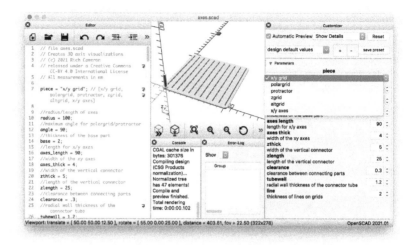

Models assumed to be using the Customizer will have a set or range of values that will then show up in a Customizer window. For example,

```
size = 2; // [2, 5, 10]
```

would show the values 2, 5, and 10 for you to pick from (defaulted to 2). Alternatively,

```
size = 2; // [2:10]
```

would give the user a slider to allow picking any integer value from 2 through 10 inclusive. If you want to have non-integer values, you can specify something like this, where the middle value gives you the increment in values you are allowed:

```
size = 2; // [2:0.0001:10]
```

Once you change anything using the Customizer, all the settings there will override the ones in the code until you close and re-open the file. For more details about how the Customizer works, check it out in the **OpenSCAD documentation** (https://en.wikibooks.org/wiki/OpenSCAD_User_Manual/Customizer).

Values in the Customizer are not visible until a model has been previewed or rendered, so do that first to get started. As with any changes you make, you'll need to re-render the model before you can create a new STL.

Downloading the Models: Github

The models reside on **Github** (https://github.com), specifically in **this book's repository** (https://github.com/whosawhatsis/Trigonometry) there. Github is a site that has many repositories, which are collections of freely available computer models, software, and so on. This repository includes the models for this book as well as a few that are used in other, overlapping projects of ours. We also plan to keep adding to it, so there may be some bonus models in there that are not described in the book. The start of each chapter has a sidebar that lists which models (and other supplies) are used in the chapter.

To download all the models, go to the link we just noted, and click the "Releases" button. There, you can download the latest version of all editable OpenSCAD files, plus select STL files. In many cases, the text will suggest you play around with a lot of variations. As a practical matter, we give you just the STL files that make representative cases, usually the prints used in the Figures. You will need to make STL files for the rest in OpenSCAD. Alternatively, you can click on just one model in the list on the repository page. It will open the OpenSCAD text file, which you can select, copy, and paste into OpenSCAD, or use the "RAW" button to download the file.

The models are released under a Creative Commons Attribution 4.0 International license. You can read more about these licenses on the **Creative Commons website** (https://creativecommons.org). This license means that you can do whatever you want with the models — print them, be paid by someone else to print them, create variations — as long as you attribute them to us. Each model has a few comment lines at the top that give the language we ask that you use for attribution.

Note that Github requires that users be at least 13 years old. If you are under 13, please ask an adult to download the files for you.

Some Models Have Small Parts

These are educational models, and some have small parts. They are intended for middle- and high-school students, and should not be used around very young children. Some models will not print correctly if they are scaled up or down. Trying to make parts smaller might make them unprintable, while increasing the size may make things fit together too loosely. As we go along, we note which models cannot be scaled from a printability point of view.

3D Printing

If you have never tried 3D printing, you might wonder how it all works. The models in this book are designed to print on a 3D printer that uses *filament* as its working material. Filament is a plastic string, typically sold in 1 kg spools. A 3D printer melts the filament and extrudes it through a fine nozzle, most commonly 0.4 mm in diameter. A 3D object is built up one layer at a time on a build platform from this extruded plastic (Figure 2-5).

FIGURE 2-5

A 3D printer, annotated.

There are many different designs of 3D printer. Some keep their build platform stationary and have the extruder move in all three dimensions. In others, the build platform moves in one dimension and the extruder head moves in two, and so on. In the case of the 3D printer in Figure 2-5 the platform moves in one horizontal dimension, the extruder moves in the other horizontal dimension along a gantry, and that whole gantry moves vertically.

Many people describe a 3D printer as a robotic hot glue gun, which is really a better analogy than a paper printer. There is no particular relationship between printers on paper and 3D printers.

3D Printing Workflow

There is also quite a bit of software involved in the 3D printing process, as well as the physical 3D printer hardware (Figure 2-6). First, a user creates a 3D CAD model (for example, in OpenSCAD). Models can be saved in several formats in OpenSCAD. Models saved as `.scad` files can be edited in OpenSCAD, but are not ready to 3D print.

For 3D printing, you have to "render" a model, which changes the model from a set of geometric shapes that have been combined in various ways into a surface covered with triangles. A file consisting of a list of these triangles and their orientation is called an *STL file*, and the process of covering a surface with triangles is called *tessellation*. A 3D model is represented in an STL file as a long, long list of triangles and information on how those triangles are oriented. You may also hear these files referred to as "meshes." The STL file is the last point in the 3D printing software chain that is printer independent. OpenSCAD takes you to this point in the process.

FIGURE 2-6

The 3D printing software flow

Once you have an STL file, the next step is to run a *slicing* program. Sometimes also called a "slicer," this program takes in an STL file and produces a file of commands for a printer to execute, one layer at a time. The slicer has a variety of settings, like the temperature of the nozzle, how thick each layer is, and so on. The slicer's output will be just for one kind of 3D printer, one material, and one CAD model.

For most 3D printers, the instructions that come from the slicer are in a format called *G-code*, though some printers use proprietary alternatives. Putting G-code into the printer often involves putting a G-code file on an SD card or USB drive and plugging it into the printer, then selecting the file from the printer's on-screen menu. Some printers also use a Wi-Fi connection to exchange files.

Since this is not primarily a book about 3D printing, we are not going to go more deeply into the process of printing. If you need more background, we suggest some books and other resources in the References section at the end of this chapter.

You may be lucky enough to have a printer at home. Many public libraries have printers available for patron use, typically after some basic training and perhaps for a small fee. If you are a student, your school might have some printers, too. *Make:* magazine routinely reviews printers if you want some advice on buying one. There are decent ones now for under $500; read reviews, and know your limitations if you are thinking about saving money with a kit. We have designed the models so they fit on the smaller end of consumer printers (ones about 150 mm in each dimension).

Materials

All the models in the book print fine in basic PLA (polylactic acid), one of the cheapest and most common 3D printing materials, which is typically made from corn or sugar cane. You can typically buy basic PLA for about $15 a kilogram, which would probably be enough to print everything in the book with some left over.

For aesthetics, we used translucent or transparent PLA for some prints, which tends to be a little pricier, or transparent PETG (polyethylene terephthalate glycol-modified, a modified formulation of the plastic commonly used for soda bottles) which can be a little harder to work with. Some models were printed with "silk PLA," which is PLA with additives that give it an almost metallic shine. It also helps to hide the layer lines and give a smoother-looking finish (Figure 2-7).

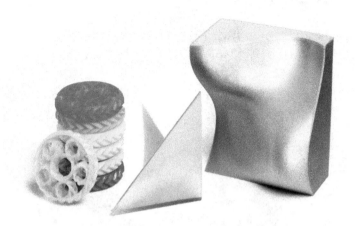

FIGURE 2-7

*Three prints (L to R):
in basic PLA, PETG,
and silk PLA*

Printing Tips

In some cases, we have found it best to "vase print" the models. Various 3D printing slicing programs might also call the setting to make this happen "spiral vase" or "spiralize outer contour." Whatever it is called, the print is created as a hollow "vase" with a solid bottom, no infill, and open top. This makes printing faster and uses less material, but only works for certain types of models.

It is always possible to print a model "on its side" so that surface detail is printed on the vertical (or near-vertical) surfaces (Figure 2-8). Printing "sideways" was used to get some of the higher-resolution surfaces, like the wave models in Chapter 8. You may need to play around with the best way to orient the models for your printer.

FIGURE 2-8

*Model from Chapter
8 printed
"sideways"*

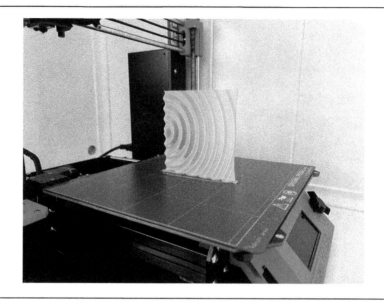

These models can be printed on a small consumer-level 3D printer and no heated bed. We generally tried to limit all the models to a maximum dimension of 100 mm to ensure that they are easy to print, even on smaller printers. (Some of the photographed models were scaled up beyond that for demonstration purposes.) Many of these models are scaled to have a minimum feature size no smaller than 1 mm.

If your printer has a nozzle larger than 0.5 mm, you might want to scale up these models a little. We do not recommend scaling most of them any smaller, since you may find some features disappear or otherwise become too small to print. In the cases where we have used features small enough for the printer's line width to matter, we have included a variable (usually called "thick" to denote the thickness of the model in the x-y plane) in the OpenSCAD source that allows you to customize the value to match your machine. As a rule, the minimum thickness of the model in the horizontal direction should be at least twice your extrusion width, and exactly four times your extrusion width is often ideal. Your slicer may default your extrusion width to somewhere between your nozzle diameter and a bit bigger than that.

You should not have to use support for most of the models (at least for the orientations and functions we used). In the few cases that are an exception, we point it out. Note that you may need to rotate some of the models in your slicing software to avoid that, though.

If You Do Not Have a 3D Printer

If you do not have a 3D printer and want to have the models somehow, you have several options. First, you can always just simulate the models in OpenSCAD. It is not as good as having the physical models, but probably an improvement on just counting on the illustrations in this book. Of course, that only works for models that do not have to fit into another piece, since OpenSCAD can only open one file at a time.

Where possible, we have also included an option to generate a 2D shape from the OpenSCAD models. These generally need to be printed on a paper printer, then cut out and folded. Models that support this behavior will have a variable that you can set to generate a 2D version. To export a 2D file, you first need to render the model, just as with a 3D file export. When the model is rendered as a 2D object, it will appear as a red outline filled with green. OpenSCAD supports both PDF and SVG formats for exporting 2D models, but the PDF option adds some scale markers and other decorations that sometimes interfere with printing. We have found that it is better to use the SVG export format. This can be opened in any web browser for paper printing, and is also a common format for use with laser cutters.

The next cheapest option is to find a friend who has a 3D printer. All the models are designed to work on a basic consumer 3D printer, for the most part without support. If your school has a makerspace, you might be able to print some of them there. It is a fair amount of printing, though, to make *all* the models for the book, so be sure you understand that when you ask. Failing that, you could send a set of the STLs to a service bureau (people who print things for others commercially) but that would be expensive.

Chapter Key Points

In this chapter, we introduced the basic mechanics of OpenSCAD and of 3D printing. We went through some of the peculiarities of OpenSCAD, relative to the programming languages like C or Java which it otherwise closely resembles. There are two general types of models for this book, which can either be edited in OpenSCAD directly or edited through the Customizer function in OpenSCAD. We gave an example of each, which also gave us the opportunity to go through other OpenSCAD structures in a bit more detail. Finally, we covered some options for readers who may not have easy access to a 3D printer.

Terminology and Symbols

Here are some terms and symbols from the chapter you can look up for more in-depth information.

- CAD, computer-aided design
- meshes
- OpenSCAD
- PETG, polyethylene terephthalate glycol-modified
- PLA, polylactic acid
- STL file
- `.scad` file

References

If 3D printing is new to you, there are various resources that might be helpful. The 2020 second edition of our book *Mastering 3D Printing* from Apress might be a good place to start. We have various training resources linked on our **website** (https://www.nonscriptum.com), including courses on the LinkedIn Learning (formerly Lynda.com) platform.

Our 2021 *Make: Geometry* book has three chapters introducing OpenSCAD in more depth, with a focus on how to use it to encourage basic geometrical reasoning. For OpenSCAD itself, we suggest **their online documentation** (https://openscad.org) as a place to start. The main developer of OpenSCAD, Marius Kintel, co-authored a book on learning to program with OpenSCAD (Gohde and Kintel, *Programming with OpenSCAD: A Beginner's Guide to Coding 3D Printable Objects*, No Starch Press, 2021).

We also suggest you look at some of the simpler examples available in OpenSCAD under `File > Examples`. If coding is also new for you, you might get started with the Gohde and Kintel book. Alternatively, look into one of the many resources on basics of coding in the C or Javascript programming languages, which are probably the closest matches.

3 Triangles and Trigonometry

3D printed models used in this chapter

See Chapter 2 for directions on where and how to download these models.

- `TriangleAngles.scad`
  - Demonstrates that the angles of a triangle sum to 180 degrees
- `ExtrudedTriangle.scad`
  - Just two lines long, this model draws a triangle from any three points and extrudes it to the desired thickness.
- `TriangleSolver.scad`
  - Creates three similar triangles (scaled). They can be specified using one of several rules, described in the text.
- `hypotenuse.scad`
  - Creates a model to explore sine and cosine

Other supplies for this chapter

- 25, 2 by 2 square LEGO bricks
- A few pieces of construction or similar heavy paper
- Regular paper to sketch on
- A calculator that can find sine, cosine, and tangent
- Ruler (one-foot works best)
- Masking tape
- Measuring tape

A *triangle* (Figure 3-1) is the simplest way you can make a shape with straight lines that has a distinct inside and outside. It is made of three lines. Mathematicians call a point where the lines meet a vertex (plural *vertices*). In this chapter, we review basic features of triangles that underlie a lot of geometry and trigonometry. Our book *Make: Geometry* explores these basics in more depth (particularly in Chapters 5 and 6) should you need a refresher. The 3D printed models in this chapter are also featured in *Make: Geometry*.

Later in the chapter, once we know what some of these shapes are called, we see why some of them are particularly useful. For starters, we try out ways of measuring things that are too big to measure directly, just by using some of the ideas in this chapter and taking a few easier-to-get measurements.

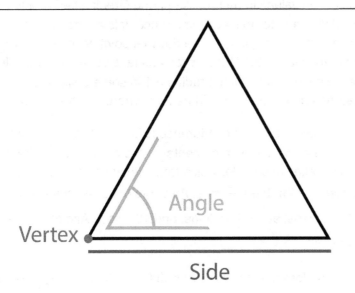

FIGURE 3-1

The anatomy of a triangle

Angles of a Triangle

Triangles with one angle equal to 90 degrees are called *right triangles,* and a little later in this chapter we learn more about special ratios of their angles and sides. As one angle of a triangle gets narrower, the triangle pinches together at that vertex. At the same time, the other two angles need to get correspondingly bigger so that it stays a triangle. We can either use the 3D printed model `TriangleAngles.scad` or just a triangle cut out of paper to demonstrate that, as it turns out, the angles inside a triangle always add up to 180 degrees. To do that, we use an old trick that dates back to Euclid (who lived around 2400 years ago in Greece).

Degrees, Radians, and Pi

Angles can be measured in degrees (symbol °). A full circle is 360°, and a rotation of 180° points us in the opposite direction (half a circle). The number 360 was chosen (most likely by the ancient Babylonians) because it is easy to divide into halves, thirds, fifths, etc.

Sometimes it is more useful to be able to think about fractions of a circle in terms of the portion of the circle's perimeter (or circumference) you have traversed. The unit of that measurement is the *radian*, which is based solely on values in nature, like pi (the Greek letter π). π is the ratio of any circle's diameter to its circumference. It is an irrational number, which means that its digits after the decimal point never end or repeat, but it is approximately 3.14159. 2π radians take us around a whole circle. Imagine we have a circle with a radius = 1 in some units. The circumference of that circle would be $2\pi r$, where r is the radius.

Since the radius is 1, the circumference is just 2π. If we had a wedge of the circle with a 45° angle at the center of the circle, it would have gone through 45/360ths of a circle, which translates to 1/8th of a circle. If the circumference is 2π, then $\frac{2\pi}{8} = \frac{\pi}{4}$. An angle of 45° is thus the same as an angle of $\frac{\pi}{4}$ radians. 90° is $\frac{\pi}{2}$ radians, and 180° is π. And of course, 360° is 2π radians.

In general, the formula to convert is $\text{radians} = \text{degrees} * \frac{2\pi}{360°}$, or, more simply, $\text{radians} = \text{degrees} * \frac{\pi}{180°}$. Converting the other way would give us $\text{degrees} = \text{radians} * \frac{180°}{\pi}$.

If you use a calculator or computer program to compute the ratios in this chapter and beyond, be sure you know whether it is using degrees or radians. OpenSCAD uses degrees, but most other computer languages use radians. Calculators that include trigonometric functions usually let you switch between degree and radian modes. Google's calculator uses radians.

Just to make things more interesting, sometimes degrees are shown as decimal degrees, like 34.1028°. Other times, they are shown in degrees, minutes, and seconds (written like this: 34° 6' 10", which I would read 34 degrees, 6 minutes, 10 seconds).

To go from degrees-minutes-seconds to fractional degrees, divide minutes by 60 and seconds by 3600, and add the result to the number of degrees. Example: 34° 6' 10" is $34 + \frac{6}{60} + \frac{10}{3600} = 34.1028°$.

To go the other way (from decimal degrees to minutes and seconds) multiply the decimal part by 60 to get minutes. For example, if we have 34.1028° and multiply the 0.1 by 60, that is 6.1667 minutes. If there was a fractional minute, we would multiply that fraction by 60 again, in this case, 0.1667 times 60, or 10 seconds.

Prove Triangle Angles Add to 180°

Make a triangle cut into four pieces, either by 3D printing the `TriangleAngles.scad` model or using paper. To make a paper version, first cut out any triangle you like. Then, using a compass, cut off each of the triangle's vertices (as we show with the 3D printed triangle in (Figure 3-2)). You need to keep track of which corner of each piece was a vertex of the original triangle, and making a curved cut as we have done here makes that easier.

FIGURE 3-2

The triangle assembled

Now remove these three vertex pieces from the triangle and put the three vertices together, so they make a straight line, as well as half a circle (and thus add up to 180°). You can see this in Figure 3-3.

The longest side has the biggest angle directly across from it, and the smallest side has the smallest angle across from it. Of course, that means that if two angles are equal, the sides opposite them are the same length. Print, or cut out, several variations of this model (changing the angles each time) to get some intuition about why this is true. If some of the angles get wider, the others have to become correspondingly smaller so that the sum of the angles remains the same.

FIGURE 3-3

The triangle's vertices rearranged to show that the angles sum up to a straight line.

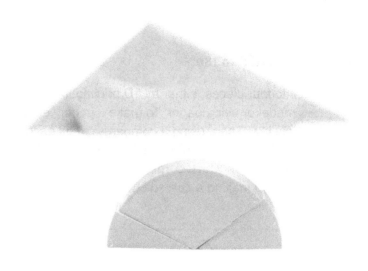

For right triangles (with, by definition, one 90° angle) this means that the other two angles must add up to 180° - 90°, or 90°. Since each of these other angles has to be less than 90°, the side that does not touch the 90° angle is the longest. This longest side of a right triangle (the one opposite the right angle) is called the *hypotenuse*. The longest side (opposite the biggest angle) does not have a special name in other triangles.

How to Use a Protractor

A protractor is a device for measuring angles. You can print one out (search online for "download protractor") or you can buy one. If you buy one, a clear plastic one is handy because it is easier to see what you are doing.

If you are measuring the angles of a triangle with a protractor, first put the vertex of the angle you are measuring on the crosshairs at the bottom of the protractor. Line up the bottom of the angle with the line on the bottom of the protractor (Figure 3-4).

Then you can read off the angle from the scale around the edge. Either read up from the bottom if the angle is less than 90° (as in this case, where the angle is about 47°) or the outer scale if it is more than 90°. Note that how precisely we can measure comes down to how good our tools are. The width of the lines making up your triangle, how accurately your protractor is printed, and how good you are at estimating all come into play. A plastic protractor like the one shown is probably good to plus or minus half a degree.

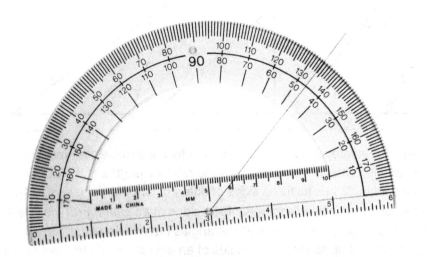

FIGURE 3-4

Measuring an angle with a protractor. (Base of the angle is parallel to the line across the bottom of the protractor.)

Drawing and Labeling Triangles

It can be convenient to note whether the sides and angles of a triangle are equal to each other. Often in geometry, we do not care what the actual measurement is; we just want to know which sides or angles are equal. Different authors use different conventions. A very common one that we use is to show one, two, or three little hash marks on the side of a triangle. If two sides are the same, they have the same number of marks. The angle across from that side gets the same number of little arcs. Let's use this convention to introduce different types of triangles.

A *right* triangle has one 90° angle, shown by using a square instead of a circular arc. If you were to widen out the right angle, you would see that the two sides making that angle would splay apart, and the opposite side would be longer than the hypotenuse was when it was a right triangle. This is now an *obtuse* triangle, with one angle more than 90°. If we were to narrow down the 90° angle, we would get an *acute* triangle. Acute triangles have all angles less than 90° (Figure 3-5).

FIGURE 3-5

Acute, right, and obtuse triangles.

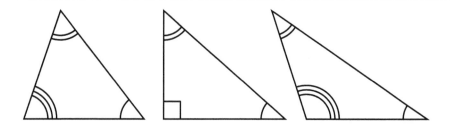

These three categories were determined by the triangle's largest angle. We can also categorize triangles by the relative lengths of their sides. A *scalene* triangle has all three sides of different lengths. An *isosceles* has two sides of the same length. All three sides of an *equilateral* triangle are the same length. That means all the angles have to be equal, too. Since they must add up to 180°, the angles of an equilateral triangle are all 60° (Figure 3-6).

FIGURE 3-6

Scalene, isosceles, and equilateral triangles

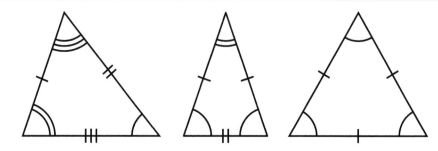

If two sides of a triangle are the same length as each other, the two angles opposite those sides are the same, too. We could also think of equilateral triangles as having all three angles the same, isosceles having two angles the same, and scalene as having all angles different.

Earlier in this chapter, we saw that the angles of a triangle add up to 180°. That means that the remaining two angles of a right triangle would add up to 180° - 90° = 90°. This tells us that right triangles cannot be acute or obtuse. The definitions of acute and obtuse mean that a triangle cannot be both, and so we have three categories that do not overlap.

In the same way, the definitions of equilateral, isosceles, and scalene do not overlap. Some people like to say that isosceles triangles have *at least* two sides the same, and consider equilateral triangles (where all the sides are equal) a special case of isosceles, but we think that is confusing and will not use the words that way.

Sometimes we may label an angle or a side with letters or their numerical values. You may also see some books that draw a triangle like the one in Figure 3-7, and then refer to angle *ABC*. Some books would call the angle in Figure 3-7 "angle B". The bottom line is that you will need to check what the book or website you are using is doing. We use the convention of little hatch marks and arcs, as in Figure 3-6, except for a few places where it is clearer to do it the Figure 3-7 way.

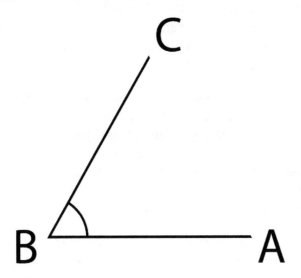

FIGURE 3-7

How to show a labeled angle

Congruent Triangles

Congruent triangles are the same size and have the same angles as each other (although they can be oriented differently). If we 3D print a triangle or cut one out of a piece of paper, it is obvious that if we take this plastic or paper triangle and flip it over (mirror it), rotate it around its center, or move (which mathematicians would call "translating"), it the triangle does not change. If we were to do any of those things and trace around the triangle in the starting position and then trace around the triangle after we had rotated, translated, and/or mirrored it, each of those traced triangles would be *congruent* to the one we started with.

You cannot stretch the triangle and have it stay congruent, though. The congruent triangles are all ones that could be drawn around the same physical triangle, which means all the sides and angles of the triangle must stay constant.

However, you do not need to know all three sides and all three angles to say that two triangles are congruent. How many angles and/or sides of a triangle do you need to know to pin down all its dimensions and angles? Do a few sketches to try to figure it out before you read on.

Okay, let's see how close you were. It turns out that if you know two angles of a triangle, you know all three (because they must add up to 180°, as we learned in the last section). When you set the three angles of a triangle, all the triangles you can make with those angles will be *similar* triangles. They will be the same shape, but because none of the lengths have been provided they will not necessarily be the same size (and therefore not congruent).

However, if you know the lengths of two sides and the angle between them (the "included angle"), that is enough to specify the whole triangle. This is known as the "side-angle-side" (or SAS) method of seeing if two triangles are congruent. Knowing all three sides will do it too (SSS) since that completely determines the possible angles of the triangle. Knowing two angles and their included side (ASA) is another option, as well as one side, one angle adjacent to it, and the opposite angle (SAA).

However, some combinations do not work. Just having all the angles ("angle-angle-angle", or AAA) determines the triangle's shape, but not its size. SSA (two sides and the angle that is NOT between them) does not always define a triangle uniquely, as we can see in the example in Figure 3-8. With the angle and the side on the left (the one with one hash mark), and a second side the length of the two equal-length lines (shown with two hash marks), we can construct two different triangles that fit the criteria. Obviously, they are different triangles, even though they both have an angle in common and two sides of the same length.

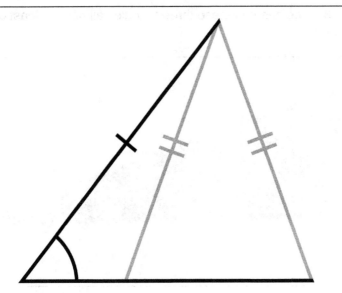

FIGURE 3-8

Showing two different triangles can be constructed for a given side–side–angle

Depending on the values, one of these two possible triangles shown in Figure 3-8 might not be a valid triangle, in which case you may be able to define the triangle uniquely from an SSA definition, but that is not true generally. There are also SSA combinations that do not make any valid triangles at all.

Similar Triangles

If two triangles have two angles the same, the third angle must be the same as well. The triangles might be different sizes, but regardless of the size, the *ratios* of the sides are the same. These are called *similar* triangles. We use an OpenSCAD model and 3D prints here, but you could also construct these triangles using paper and a protractor instead.

The OpenSCAD model `TriangleSolver.scad` creates three similar triangles. You need to specify one triangle and then a set of scaling factors is applied to create the desired number of triangles. Another quick way to prove that the three angles are the same in all three triangles is to overlay them, with the relevant angle lined up (Figure 3-9).

Note that similar triangles have the same set of angles, but are not (necessarily) congruent because they may be different sizes, as we see here. If the sizes of the triangles were the same, they would indeed be congruent, so congruent triangles are a special case of similar triangles.

In Figure 3-9, we can see these three triangles all have at least one angle in common.

FIGURE 3-9

Similar triangles lined up to show their angles are the same.

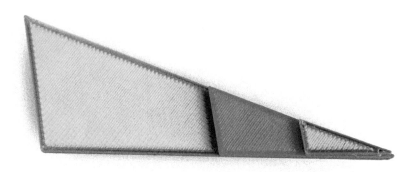

Mixing the three scaled triangles still allows you to create a straight line out of the angles, just as we saw in the exercise where we lined up the three angles of one triangle earlier in the chapter. In other words, the angles add up to 180° even if you use the angles from the same triangle scaled up or down.

To test this out for yourself, take a set of similar triangles created by `TriangleSolver.scad` (or that you create with paper and a protractor) and show that even though the scales of the triangles are different, all three angles are the same (Figure 3-10). We took three similar triangles (which each have the same three angles), and we turned each triangle so that we used one different angle from each one. So, it is equivalent to cutting off the three angles of one triangle and aligning them, as we did in Figures 3-2 and 3-3.

FIGURE 3-10

The three similar triangles in Figure 3–9 rearranged to show that taking one angle from each triangle still adds up to 180°

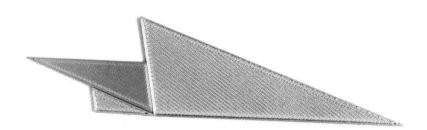

Trigonometry Basics

No parts of mathematics are really completely separate from the others. In particular, geometry and trigonometry have some common roots that we need to comprehend to understand both subjects. In this section, we introduce a bit of trigonometry that describes useful relationships between the sides and angles of right triangles. First, we define these relationships, and then we put them through their paces and see how they can help us out.

Pythagorean Theorem

About 2500 years ago, Pythagoras (whom we first met in Chapter 1) figured out that if you have a triangle that has one right angle, there is a relationship between the longest side and the two shorter ones. In other words, if we add the squares of the two shorter sides, we get the square of the longest side (the hypotenuse). As an equation, if the two shorter sides are of length a and b and the longest is of length c, the Pythagorean Theorem says

$$a^2 + b^2 = c^2$$

For example, if the hypotenuse of a triangle is 30 mm long, then 30 mm squared, or 900 sq mm, is the sum of the squares of the sides. Let's say that we want the other two sides to be the same length as each other ($a = b$). We know that

$$a^2 + b^2 = c^2$$

Since $a = b$, then $2a^2 = c^2$, and so $2a^2 = 900\text{mm}^2$ (note that units have to be squared, too).

Dividing both sides by 2,

$$a^2 = 900/2\text{mm}^2$$

Now take the square root of both sides, to get our answer:

$$a = \sqrt{450\text{mm}^2} = 21.2\text{mm}$$

Squares and Square Roots

If symbols like a^2 are new to you, they may look scary but are easy to understand. What a^2 means is "take the number a and multiply it by itself," which is also called "squaring" a number. Or, in other words, $a^2 = a * a$. Then if we want to multiply by a one more time, we write $a^3 = a * a * a$, and so on.

We can go the other direction, too. Suppose we know that $a^2 = 2$. That means that something times itself equals 2. Figuring this out is called taking a *square root*, and its symbol looks like this: $\sqrt{2}$. You probably have a button on your calculator that says "$\sqrt{}$", and if you want to use Google to get the square root of 2, you would type `sqrt(2)`.

Note that $\sqrt{4} = 2$. But also, if we multiply -2 times -2, we get 4. The positive result is called the principal value, and what we will use as a default. But the other option is valid (and sometimes what we want). See Chapter 4 for more on principal values. The square root of a negative number involves something called "imaginary numbers," and you can learn about those in our *Make: Calculus* book.

Pythagoras' LEGO Bricks

We can demonstrate the Pythagorean Theorem with LEGO bricks and a 3D printed triangle. The dimensions of a 2 by 2 LEGO brick are 16 mm by 16 mm by just shy of 10 mm deep. If we have four of them on one side (64 mm), and three on the other (48 mm), that means that the square of the longest side should be 3 squared plus 4 squared, or $9 + 16 = 25$.

You can use the model `ExtrudedTriangle.scad` to generate a triangle with one side 64 mm long and the other 48 mm long, at right angles to each other. We want it to be 10 mm thick. To do that, update `ExtrudedTriangle.scad` as follows. Or just use these two lines as your model; see Chapter 2 to learn how to make a new OpenSCAD model.

```
thickness = 10;
linear_extrude(thickness) polygon([[0, 0], [64, 0], [64, 48]]);
```

Take that triangle and 25, 2 by 2 LEGO bricks. Make a square of nine of the bricks and a square of 16 of them. These align with the two short sides (Figure 3-11). Now rearrange these bricks into a 5 by 5 square, which lines up with the hypotenuse (Figure 3-12).

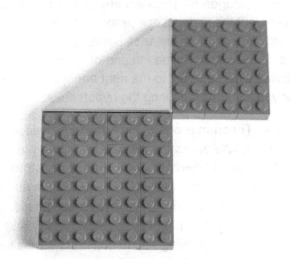

FIGURE 3-11

The sides of the triangle line up with 3 and 4 bricks each.

FIGURE 3-12

The hypotenuse of the triangle lines up with 5 bricks.

If you use a different-size counting piece, you will need to adjust the triangle you create. There are a few "Pythagorean Triples" that come out to nice round numbers like this (the next one is 5, 12, 13), and there are long lists available if you search on that phrase. Alternatively, you can read the Wikipedia article "Pythagorean Triples" to see some history and formulas to generate as many as you like.

You can use the Pythagorean Theorem any time it is easy to measure two distances or lengths in directions perpendicular to each other, and you want to know the shortest distance between the farthest-apart points of the triangle. If you started at the left end of the hypotenuse and traveled right and then up to get to the right end, your path would be longer than if you just traveled along the hypotenuse.

The Pythagorean Theorem is one of the best-known pieces of mathematics, and a little searching will give you more proofs if you would like to see it approached differently. A related fun construction is called the "Spiral of Theodorus," and we explore it in depth in our *Make: Geometry* book. It is a construction of successive right triangles which have hypotenuses of square root of 2, square root of 3, and so on (Figure 3-13). This amounts to an analog way to construct values for square roots of numbers up to 17. Figure 3-14 is a 3D printed version of the construction up to 17; directions for creating it are in *Make: Geometry*.

FIGURE 3-13

The first two triangles in a Spiral of Theodorus

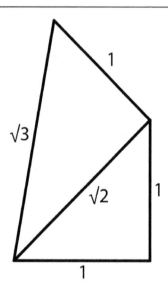

This construction can, in principle, be used to measure the square root of any integer, but as you can see, beyond $\sqrt{17}$, the triangles will start to overlap.

FIGURE 3-14

A Spiral of Theodorus showing hypotenuses up to square root of 17

Sine, Cosine, Tangent

It turns out that there are ratios among the sides of right triangles that can be used to calculate angles, or other sides of the triangles. Three ratios are used most commonly, called *sine*, *cosine*, and *tangent*. These deal with the ratios of the triangle's three sides. The hypotenuse is one of these sides, and the other two are referred to as the *adjacent* side (the side sharing a particular angle with the hypotenuse) and the *opposite* side (the third side of the triangle, which does not share that angle).

The *sine* of an angle a (usually written sin(a)) in a right triangle is

$$\sin(a) = \frac{\text{opposite}}{\text{hypotenuse}}$$

The *cosine* of an angle a (usually written cos(a)) is

$$\cos(a) = \frac{\text{adjacent}}{\text{hypotenuse}}$$

The *tangent* of an angle a (usually written tan(a)) is

$$\tan(a) = \frac{\text{opposite}}{\text{adjacent}} = \frac{\sin(a)}{\cos(a)}$$

Hypotenuse Model

Let's use the model `hypotenuse.scad` to get some intuition about these ratios. First, assemble the model. It has two legs; one has a slot cut through it, and one has a slot cut part-way through and is solid at the base. Place the model so that the leg that is solid at the base is on the right side, and the one with the slot all the way through is on the bottom. It will look like a backward letter L. Next, take the slider and put one of its tabs into each slot. The slider plus the L-shaped part makes a right triangle (Figure 3-15). You now have an analog calculator of sorts for sine and cosine.

FIGURE 3-15

The hypotenuse model assembled

Let's talk about the angle on the bottom left of the triangle first. Assume that the hypotenuse of the triangle made by the slider is one unit long. The bottom and side of the L shape are marked off in tenths and hundredths of this "unit hypotenuse" (Figure 3-16).

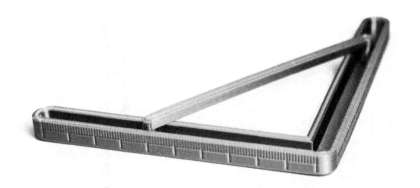

FIGURE 3-16

Labeling on the bottom and side of the model

Therefore, since sine is equal to $\frac{\text{opposite}}{\text{hypotenuse}}$, we can read off the sine of this angle on the right side, and cosine off the bottom (adjacent) side. The zero point is at the lower right (where the two legs join) and the tenths are marked with long lines, with nine shorter lines in-between for reading hundredths. Thus, you should be able to read the value of the sine or cosine to two digits. The lines are indented for better printing and longer model life.

Similarly, you can read off the cosine of the angle from the scale along the bottom of the model (the side cut all the way through). Read it *right to left* — that is, zero is at the corner for *both* scales. Figure 3-16 shows a cosine reading of about 0.62.

Since the hypotenuse is the longest side of the triangle, the value of sine goes from 0 to 1. You can see this by sliding the slider all the way over one way till it is nearly flat (near 0°, as shown in Figure 3-17) to the other end where it is nearly vertical (90°, as shown in Figure 3-18).

FIGURE 3-17

0° shown on the model

FIGURE 3-18

90° shown on the model

Finally, what about the tangent of this angle? Tangent is the opposite side over the adjacent, or sine divided by cosine. This means it is how much the slider rises from left to right versus how far it is from the corner. Unlike sine and cosine, there is no limit to the value of the tangent of an angle. As the angle approaches 90°, the tangent approaches infinity, because the bottom of the fraction $\frac{\text{opposite}}{\text{adjacent}}$ is approaching zero. Two angles of a triangle cannot both be 90°, though, since the angles need to add up to

180°. That is consistent with the fact that dividing by zero is also undefined.

Test this out. Use the model to find what angle gives you a sine of about 0.71. What is the cosine of this angle? Describe what happens to the sine, cosine, and tangent of an angle when the angle gets close to zero. What about as it gets close to 90°? Use the model to help demonstrate and think about this. (See the answers at the end of the chapter for the numbers.)

Complementary Angles

We know that the other two angles in a right triangle add up to 90°. If we use the model and think about the two angles, the sine of one angle is the cosine of the other. The "co" in cosine stands for "complementary." The sine of an angle is the same as the cosine of its complementary angle. Or to put it another way,

$$\sin (x) = \cos (90° - x)$$
$$\cos (x) = \sin (90° - x)$$

Along with complementary angles, you may hear the term *supplementary* angles. This refers to two angles that add up to 180°. A pair of supplementary angles have the same sine, and their cosines add up to zero, as we will see when we talk about angles between 90° and 180° in Chapter 5.

Other Ratios

Sometimes it is more convenient to use other trigonometric ratios, defined as follows:

$$\text{Secant: } \sec (x) = \frac{1}{\cos (x)}$$

$$\text{Cosecant: } \csc (x) = \frac{1}{\sin (x)}$$

$$\text{Cotangent: } \cot (x) = \frac{1}{\tan (x)}$$

Like sine and cosine, the other complementary (co-) ratios have this relationship with the complementary angles:

$$\tan (x) = \cot (90° - x)$$

$$\sec(x) = \csc(90° - x)$$

One tricky thing, though, is that because cosine, sine, and tangent can all go to zero, secant, cosecant, and cotangent will approach infinity when those numbers in the denominator go to zero. (A mathematician would say that dividing by zero is *undefined*, and these ratios are said to be *indeterminate* as they approach infinity.)

Angles Greater Than 90°

The trigonometric ratios have a broader definition that moves beyond just the relationships among angles of a right triangle. Chapter 5 introduces this broader definition in terms of a construct called the *unit circle*, with the help of more 3D printable models. There are also a lot of handy relationships among these (called "trig identities") which we explore in Chapter 6.

Finding the Length of a Side

Let's suppose that we know that an angle of a right triangle is 30°, and the hypotenuse is 5 cm long. What is the length of the opposite side? We can use a calculator (or estimate with our model) to find out that the sine of 30° is 0.5. This means that the opposite side is 0.5 times as long as the hypotenuse, or 2.5 cm:

$$\sin(30°) = 0.5 = \frac{\text{opposite}}{\text{hypotenuse}} = \frac{\text{opposite}}{5\text{cm}}$$

$$\text{opposite} = 0.5 * \text{hypotenuse} = 0.5 * 5 = 2.5$$

Do the same thing with cosine to find the adjacent side:

$$\cos(30°) = 0.866 = \frac{\text{adjacent}}{\text{hypotenuse}} = \frac{\text{adjacent}}{5\text{cm}}$$

$$\text{adjacent} = 0.866 * \text{hypotenuse} = 0.866 * 5 = 4.33$$

Note that you can check that you are right because the Pythagorean Theorem says that the squares of the two sides should equal the square of the hypotenuse. In our case here, it is true that:

$$2.5^2 + 4.33^2 = 5^2$$

$$6.25 + 18.75 = 25$$

$$25 = 25$$

so our answer is correct.

Arcsin, Arccos, Arctan

Suppose we wanted to "go backwards" and, given a sine, cosine, tangent or other ratio, find out what angle corresponds to it. In the case where the sine of an angle equals a value we will call x, we can talk about the arcsine (written asin(x), or sometimes arcsin(x) or inverse sine (written $\sin^{-1}(x)$). The other ratios have inverses named similarly.

For example $\sin(30°) = 0.5$, so $\sin^{-1}(0.5) = 30°$; $\cos(60°) = 0.5$, so $\cos^{-1}(0.5) = 60°$ and $\tan(26.6°) = 0.5$, so $\tan^{-1}(0.5) = 26.6°$. In Chapter 5, we talk about how this works for angles greater than 90°, where it gets a little complicated. We use the notation $\sin^{-1}(...)$, $\cos^{-1}(...)$, etc. in the rest of the book. In programming code, the functions would be `asin()`, `acos()`, and `atan()`.

Calculating with Sine and Cosine

You can test out how well you understood the discussion above by trying to do some calculations yourself. The answers are at the end of the chapter.

- I have an angle of 45° in a triangle with a hypotenuse 5 cm long. What is the length of the opposite side?
- What is the length of the adjacent side? Why? (Hint: If one angle is 45°, what is the other angle in a right triangle?)
- Check to see that you are right by using the Pythagorean Theorem.

Also try going back and forth between sin(30°) and $\sin^{-1}(0.5)$. Note that on some calculators, you need to hit another button (sort of like a shift key) to get to these functions.

Measure Something Big

We can exploit the properties of similar triangles to measure things that are too big to measure directly. We do need some measurements (and to understand some sources of error) to get a reasonable approximation. The closer something is to you, the bigger it appears. We can use this property and a little mathematics to determine either how big or how distant something large or far away is, if the size and distance from us are known for a smaller, closer object.

As a first example, we measure the height of the top of a doorframe indirectly. We can easily just measure this height with a tape measure (80 inches, in this case). However, if we first try out a technique when we know the answer we should get, we can use it with more confidence in cases where that is not possible. Our second example (measuring the high point on a roof) is a bit more challenging and shows the power of the technique for more-useful cases.

Testing with a Known Object

First, we try a very simple method to measure an object of unknown height, but a known distance away from us. We hold out a ruler in front of our eyes and look past the ruler to the object we are trying to measure. We can get the height of the large object by moving the ruler nearer to and farther from our eyes (while being careful to keep the ruler vertical) until the height of the ruler and the height of the object appear the same.

This works because we have constructed a pair of similar triangles. As we can see in Figure 3-19, the red triangle and the smaller, overlapping purple one are the same shape. All their corresponding angles are the same, so the ratios of their sides must also be the same. That is

$$\frac{h1}{d1} = \frac{h2}{d2}$$

The height $h2$ is known because it is a ruler. Thus, if we can also measure $d1$ and $d2$, we can solve for $h1$. This takes a few steps, though.

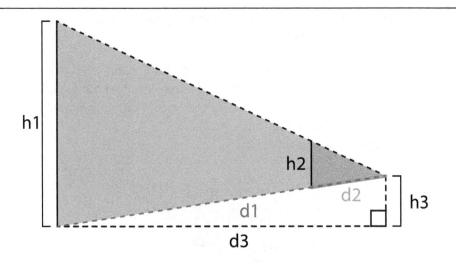

FIGURE 3-19

Layout of our measurement project

We can measure $d2$ simply with a piece of string running from between our eyes to the bottom of the ruler. We could, in principle, measure $d1$ the same way. However, a string that long would be likely to sag, giving us an answer that is too big. Instead, we create a third triangle by measuring $d3$ along the ground, then $h3$ up to our eye level. Assuming the ground is level, this is a right triangle, and its hypotenuse is of length $d1$.

$$(d1)^2 = (d3)^2 + (h3)^2$$

Next, we need to figure out how far above the floor your eyes are. Stand a foot or so from a wall. Put a small piece of masking tape on the wall at your eye level. Measure the position of the tape to get the distance between your eyes and the ground, which is $h3$ in Figure 3-19.

Then, put a piece of paper or a bit of tape on the floor to show where you will stand to make your estimate. (Try to stand so that the mark is roughly directly below the bridge of your nose.) Use a tape measure to find the distance along the ground between that mark and the object you are measuring. This is the dimension marked $d3$ in Figure 3-19. We are assuming that the floor is pretty level. If the ground slopes significantly, that will introduce errors. Consider how to re-draw (and analyze) Figure 3-19 for your situation.

Now comes the tricky bit. Cut a piece of string a bit longer than your arm and attach it to one end of the ruler. Next, pick up the ruler and hold it vertically, with the string attached to the bottom. Move the ruler closer and farther from you until it just exactly covers the larger object. Hold the ruler at that distance with one hand. To make it more concrete, let's say you are holding it with your left hand.

Without moving the ruler, take the loose end of the string in your right hand, and use the string to record the distance from the bottom of the ruler to where the bridge of a pair of glasses would be, between your eyes and level with them (Figures 3-20 and 3-21). Measure this length of string to get the distance $d2$ in Figure 3-19.

FIGURE 3-20

Lining up the ruler and doorframe

FIGURE 3-21

The setup in Figure 3-20 as seen by a third-party observer

Now we have all the measurements we need! We can find out $d1$, the distance from our eyes to the bottom of the object we are measuring, with the Pythagorean Theorem. It is

$$(d1)^2 = (d3)^2 + (h3)^2$$

Now, because we have similar triangles, the ratio of $h2$ to $h1$ must be the same as the ratio of $d2$ to $d1$:

$$\frac{h2}{h1} = \frac{d2}{d1}$$

or we can multiply this out to get

$$h1 = d1 * \frac{h2}{d2}$$

In the case of our test with the doorframe, the values we measured (in inches) were

$$h2 = 12$$
$$h3 = 64$$
$$d2 = 29$$
$$d3 = 191$$

This meant that our estimated value of $d1 = \sqrt{64^2 + 191^2} = 201$, and thus our value of $h1 = 201 * \frac{12}{29} = 83$ inches. That is a bit high (we measured it at a bit over 80 inches) but it is only about 4% off. Given that we were only measuring to the nearest inch throughout, it is not a bad estimate.

This all seems a little overkill for something this simple, but it lets us check our process. The real power of this method is that you can use it to determine the height of things way too tall to measure directly, like the height of a tree or of a garage roof.

Garage Roof Height

First, a safety note: Never sight something near the Sun! Always arrange your measurements so that the Sun is at your back, or wait a while until the Sun is farther from what you are looking at. Once you have done that, pick a spot to stand on the ground and mark it (with a piece of tape, or pebble). Write down the distance to the object you want to measure.

Then proceed as we did for the doorframe. We tried this with the peak of a garage roof (Figure 3-22).

FIGURE 3-22

The setup for measuring the height of the garage

We stood at $d3$ = 191 inches from the garage, the same as we did inside for the doorframe test, and our values for $h2$ and $h3$ also remained the same. Thus, our value of $d1$ was again 201 inches. We measured the length of the string ($d2$) to be 15 inches. Thus, the height of the peak of the garage roof was estimated to be 161 inches. We measured it (also not terribly accurately) with a very long tape measure and got 155 inches, so again about 4% difference between two rough measurements. Not bad for a ruler and string!

Chapter 7 will explain how to do more measurements like these and discuss how surveying instruments work. We will also try out a measurement of how far away the Moon is, and a way to find your latitude from the Pole Star (or Southern Hemisphere equivalents).

Chapter Key Points

Triangles are the simplest closed polygon, and as such they underlie many more complex shapes. In this chapter we learned about various categories of triangles and their properties. Next, we learned about the Pythagorean Theorem and the basic trigonometric ratios. Finally, we applied this knowledge to measure large objects that might be tricky to measure directly.

In the next chapter, we are going to talk about how to measure where we are in 2D or 3D space relative to another point. To do that, we use *coordinate systems*. There are several different ways to do that, and Chapter 4 helps you create models of three common ones and experiment with them.

Terminology and Symbols

Here are some terms from the chapter you can look up for more in-depth information.

- complementary
- cosecant, csc(a)
- cosine, cos(a)
- cotangent, cot(a)
- congruent
- degrees
- hypotenuse
- radians
- pi, π
- secant, sec(a)
- similar
- sine, sin(a)
- square root, $\sqrt{a}$
- tangent, tan(a)
- protractor
- Pythagorean Theorem

References

To follow up on this trigonometry introduction, check out the "Trigonometric Functions" entry in Wikipedia. For more on sine (note that cosine and tangent only have links that redirect to the "Trigonometric Functions" Wikipedia entry) see the "Sine" entry. The Khan Academy has many videos in this space as well. We have just scratched the surface here of the applications of trigonometry in fields like astronomy, trigonometry, and surveying. Chapter 7 has more in-depth applications to try out.

Answers

Here are the answers for the activities in this chapter that we do not solve in the text of the section.

Hypotenuse Model

- Use the model to find what angle has a value of sine of about 0.71.
  - Answer: 45°
- What is the cosine of this angle?
  - Answer: also 45°, since if one angle of a right triangle is 45° the other one has to be, too.
- Describe what happens to the sine, cosine, and tangent of an angle when the angle gets close to zero.
  - Answer: Sine approaches zero, too (as the opposite side shrinks). Cosine approaches 1 as the adjacent side becomes about the same as the hypotenuse. Tangent approaches zero, since it is sine divided by cosine.
- What about as it gets close to 90°?
  - Answer: Sine approaches 1 (as the opposite side and hypotenuse approach being equal lengths to each other). Cosine approaches 0 as the adjacent side vanishes. Tangent approaches infinity, since it is sine divided by cosine.

Calculating with Sine and Cosine

- I have an angle of 45° in a triangle with a hypotenuse 5 cm long. Length of opposite side = sin(45°) * 5 cm = 0.7071 * 5 cm = 3.54 cm.
- What is the length of the adjacent side? Also 3.54 cm.
- Check to see that you are right by using the Pythagorean Theorem.
  - $3.54^2 + 3.54^2 = 25 = 5^2$, and so it works.

4

Coordinate Systems and Analytic Geometry

3D printed models used in this chapter

See Chapter 2 for directions on where and how to download these models.

- `axes.scad`
  - This model creates various coordinate planes, and a set of coordinate axes.
- `helicoid.scad`
  - This model creates a helicoid shape sized to the polar coordinate model.
- `r(theta).scad`
  - This model creates an Archimedean spiral sized to the polar coordinate model.

You will also need

- Some aluminum foil
- A pipe cleaner (sometimes called a "chenille straw")

One of the most familiar things in any math textbook is a grid with arrows marking the direction of increasing x, y, or some other variable. We saw three-dimensional examples in Chapter 2 when we introduced OpenSCAD. However, like every other math tool we use, someone had to draw it for the first time, and then come up with ways of defining the rules of that particular road.

A coordinate system defines one or more dimensions and allows us to give coordinates for one position relative to another. For example, in the familiar grid, an axis (usually shown as a line ending in an arrow) represents a variable, x, in the horizontal direction, and another represents the variable y in the vertical direction. In that system, we might say that a point is at (3, 4). This is shorthand for saying that it is the only point where both $x = 3$ and $y = 4$ (Figure 4-1). This type of graph uses a *Cartesian coordinate system.*

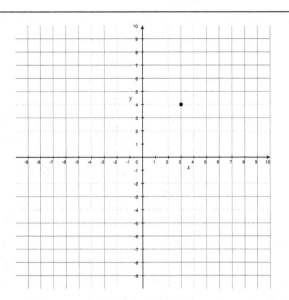

FIGURE 4-1

*A Cartesian graph of
a point at (3, 4)*

In this chapter, we introduce 3D printable models of several coordinate systems. Our *Make: Calculus* book also covers some of the territory in this chapter with a different set of models. Those emphasize area and volume elements in these different coordinate systems. The ones in this book are intended to think about locating points, curves, and surfaces in 2D and 3D space.

Analytic Geometry

Coordinate systems link algebra and geometry and, for two or more dimensional systems, allow us to talk about a variable in terms of one or more others. Legend has it that French philosopher René Descartes came up with the idea of coordinates in the early 1600s when he saw a fly wandering around on his window. The window was made up of small panes, and thus he made the leap to tracking the fly on a grid. Be that as it may, Cartesian coordinates (named for Descartes) came into being around that time.

Before that, mathematicians solved math problems by constructing geometric figures with a drawing compass and straightedge. Our *Make: Geometry* book has a chapter describing how to do a few classical constructions if you want to try it out.

However, this meant that if you could not construct a solution with those tools, you could not solve the problem. For example, numbers raised to powers higher than three were considered to be sort of curiosities, since constructions treated squared values as areas. Cubed values can also be treated as a volume of space — having height, width, and depth — but how would you draw something to the fifth power? There was no good way to conceptualize that with a compass in 2D or 3D space. Tying mathematics to one method of solving problems gave some intuition for nearly two thousand years, but it was not adequate for the types of problems solved by calculus, for example.

Descartes' 1637 book, *La Géométrie,* codified his ideas and marked the beginning of the field of analytic geometry, which ties together the geometric figures that originally were drawn by people using a compass and a straightedge, and their representations with equations based on coordinate systems. Isaac Newton was heavily influenced by Descartes' ideas when he developed calculus.

Today, analytic geometry might be included in a high-school curriculum, perhaps as the latter part of an Algebra II class, or placed within the fuzzy boundaries of "precalculus." We think of it more broadly as topics that sit in-between geometry and calculus and provide some of the path between the two.

This chapter, and Chapters 9, 10, and 11, focus on topics that are more analytic geometry than trigonometry, although a lot of trig is used in thinking about them. We have tied them together in this book because we think it is hard to learn one without the other. It is difficult to think about trigonometry beyond right triangles without a coordinate system to anchor the discussion. Let's look at four common coordinate systems and think about when to use each one.

Coordinate Systems

New York City's borough of Manhattan, with its numbered streets and avenues, is mostly laid out on a Cartesian grid. Its streets are oriented more or less north-south and east-west. A tourist can predict that 200 West 75th Street is farther west than 100 West 75th Street. It makes perfect sense to give directions like "three blocks east and one block south of Fifth Avenue and West 96th Street." Some streets have names instead of numbers, Central Park puts a hole in the grid, and the southern tip of the island is an exception. But the general principle applies.

Suppose instead that you were visiting one of the Hawaiian Islands. These islands are all roughly oval or circular in shape, with a tall volcano in the center effectively limiting road construction to roads that ring the shore and maybe a village here and there a little way inland. Thus, in Hawaii, it is common to give verbal directions in terms of "mauka" (toward the mountain, or inland) or "makai" (toward the sea) and the next town in the appropriate direction on the roads that circle the islands, like "mauka toward Kona Town."

This type of convenience for a given set of circumstances comes up in math situations too. New York City corresponds to 2D Cartesian coordinates, with two axes — x and y — at right angles to each other. All the OpenSCAD screenshots in Chapter 2 show 3D Cartesian coordinates, adding in the height dimension. You could think of the floor number in a building as a z axis coordinate, though the height of a particular floor number in two different buildings will only be roughly equivalent.

As we will see shortly, the Hawaiian example would be better served with polar coordinates (2D) or cylindrical coordinates (3D). Other applications, like latitude and longitude or positions of objects in the sky, might best be described in spherical coordinates. A curve through 2D or 3D space can be expressed in any coordinate system, but might be elegantly simple in one and a horrible mess in another.

Coordinate System Model

We have created a 3D printable model, `axes.scad`, to visualize these coordinate systems. This model can be used to create five different pieces, based on the value of the parameter `piece`. Different coordinate systems use different sets of pieces. The model creates one STL file at a time, which can always be combined in a slicer and printed all at once if the printer's build volume is big enough.

We assume in our descriptions that the x and y axes are in the plane of a table. The x axis arrow points to the right, and the y axis arrow points away from you. The third axis is the z axis. It is oriented toward the ceiling.

In Chapter 2, we used this model as an example of how to use the Customizer. Most of the model parameters will stay at their defaults, except to pick which value of `piece` to use. See the caveats in the following list of parameters before changing any of them.

Model Parameters

The `axes.scad` model has these parameters (lengths in mm, angles in degrees):

- Which model to print
  - `piece = ["x/y grid" (default), "polargrid", "protractor", "zgrid","altgrid", "x/y axes"];`
- radius of polar grid, or length of axes
  - `radius = 100;`
- maximum angle for polar grid/protractor
  - `angle = 90;`
- thickness of the base part
  - `base = 2;`
- length for x and y axes
  - `axes_length = 90;`
- width of the x and y axes
  - `axes_thick = 4;`
- width of the vertical connector
  - `zthick = 5;`
- length of the vertical connector
  - `zlength = 25;`
- clearance between connecting parts
  - `clearance = 0.3;`
- radial wall thickness of the connector tube
  - `tubewall = 1.2;`
- thickness of lines on grids
  - `line = 2;`

Most of these parameters normally will not change. They affect the printability and durability of the prints, as well as how they fit together. As we will see shortly, these models consist of two parts that slide together, via a vertical connector that fits inside a connector tube. The default clearance value of 0.3 mm should leave this joint loose enough to fit and turn freely even when created on a poorly calibrated printer. However, to make the pieces fit more snugly so that they hold their position, try reducing that value and reprinting the *x*-*y* axes piece until you get the desired fit.

The model's grid size is fixed at 1 centimeter. If you have a very large printer and want to print a larger version — for demonstration in front of a class, for example — you can safely scale the model up in your slicing software. Note, though, that the clearance between parts will scale up with it, so the resulting fit will be looser. Scaling down, however, is not recommended. Reducing the size will also reduce the clearance between parts, creating a tighter fit. However, it will also reduce the thickness of some features to the point that they may become too delicate, or even be omitted entirely by slicing software.

Cartesian Coordinates

Cartesian coordinates use three axes that are at right angles to each other, called x, y, and z. First, we use our models to find locations in 2D and 3D space, and then explore rotations around the z axis. Generate three different pieces to make a Cartesian model set, using

- `piece = "x/y grid"` (creates the *x*-*y* plane)
- `piece = "zgrid"` (creates the vertical, *y*-*z* plane)
- `piece = "x/y axes"` (creates a pair of axes).

To experiment with a 2D Cartesian coordinate system, take the `x-y grid` model piece, which has a grid with a piece sticking out of it. Let the point where the rod is sticking out be the (0, 0) point, otherwise known as the *origin*. Slide on the piece with the two axes, which you created with `piece = "x/y axes"`. The x axis should now be pointing to the right, and the y axis away from you. To find the point (3, 4) count three points right of the origin and four points up. (This is shown by a dot in Figure 4-2)

FIGURE 4-2

*Marking a point in
2D Cartesian
coordinates*

Graphing a Line

Now that we know how to graph a point in these coordinate systems, let's think about how to graph a straight line, like the one marked with a pipe cleaner in Figure 4-3. That line crosses the y (vertical) axis at $x = 0$ and $y = 3$. Now, if we look carefully, we can see that the line (shown with a pipe cleaner) goes up 1 box in y for every 2 boxes it goes right in x. The ratio of these (sometimes phrased as "rise over run") is the *slope* of the line. In this case, the slope is $\frac{1}{2}$.

Once we have the slope, we can write an equation for the line in terms of x and y. We know that when $x = 0$, $y = 3$. We also know that y will rise half as fast as x. Taking these together, the line in Figure 4-3 can be written as

$$y = \frac{1}{2}x + 3$$

In general, the equation for a straight line in Cartesian coordinates is written

$$y = mx + b$$

where m is the slope and b is the value of y when x is zero, or the y *intercept*. Why "m" and "b"? Descartes was French, and there is some speculation that it may have been some French word lost to time. Or, it

may have been used as a convention that just stuck (likewise the "*b*").
Doing an internet search on the question raises an assortment of
hypotheses if you want an excuse to procrastinate on your trig homework.

FIGURE 4-3

*A line intersecting
the y axis at y = 3
with a slope of 2*

Now, what happens if instead the line is sloping downward? Think of the
slope as the ratio of the differences in y and x between two points on a
line. If the value of y is falling while the value of x is increasing, the two
differences will have the opposite sign. In Figure 4-4 we see a line which
also intersects at $y = 3$, but now goes down 1 for every 2 box increase in
x. This is a slope of $-\frac{1}{2}$, and the line has the equation

$$y = -\frac{1}{2}x + 3$$

FIGURE 4-4

A line intersecting the y axis at y = 3 with a slope of $-\dfrac{1}{2}$

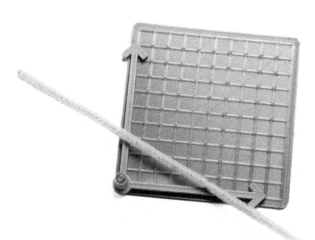

3D Coordinates

Now, let's build on this model. Leave the x-y plane where it was, but remove the piece with the arrows. Attach the second plane (made with `piece = "zgrid"`) by sliding the hollow part it has on one side over the rod at the origin of the x-y plane (Figure 4-5).

As in the previous section, find the point $x = 3$, $y = 4$. Now swing the vertical plane so that it crosses over that point. To find the point (3, 4, 5), count up 5 marks vertically (Figure 4-6).

FIGURE 4-5

*Adding the third
dimension*

FIGURE 4-6

*Finding the point (3,
4, 5)*

Note that having integer x and y coordinates will not necessarily line up
with integer lines on the z grid. Here, we were using numbers that just
happen to line up that way. The distance to that point from the origin is
$\sqrt{x^2 + y^2 + z^2}$, by the Pythagorean Theorem.

To check this, first find the distance in the x-y plane by finding the
distance $\sqrt{x^2 + y^2}$. If we create a right triangle with that x-y distance as
one side and the vertical distance as the other, the hypotenuse will be the

distance of the point from the origin. As a final check, calculate what the distance should be, then use a piece of string to measure from the origin to the relevant point in space. Measure the length of the string and see how close you are. (You will not be able to measure the distance to the origin perfectly since the pivot is occupying that space.) Play around with our pipe cleaner or some string and see what sort of shapes you get as you lay out different relationships among the variables.

If you do not have a 3D printer, you can simulate this with two pieces of graph paper. To make the second piece stand vertically, you can tape it to the side of a box, or perhaps a thick book. Unlike the 3D printed model, you will need to mark an origin point on the two pieces, and manually ensure that they are aligned with one another. Or you can use one of the many 3D graphing programs out there to have a virtual set of coordinates and visualize your points and curves.

In Chapter 8, we will explore what happens if we want to rotate this coordinate system if that is more convenient for calculations. Doing that uses some of the ideas in this section, and mashes them up a bit with what we see in the next.

Polar Coordinates

Now, let's look at other coordinate systems that might be more convenient if you are trying to solve a problem that is more naturally expressed in circles rather than right-angle grids. For example, the Hawaiian street system we described earlier, is more or less a *polar coordinate system,* in which we define position as the radial distance (usually called r) from a defined central origin, and angle (usually called *theta*) from a defined line (Figure 4-7).

Lines of constant radius are concentric circles. Lines of constant angle (usually denoted by the Greek letter theta, θ) are then spokes out from this origin. If you are right at the origin (sort of like being in the volcano in Hawaii) it is a little awkward to think about what the angle means.

FIGURE 4-7

Polar coordinates

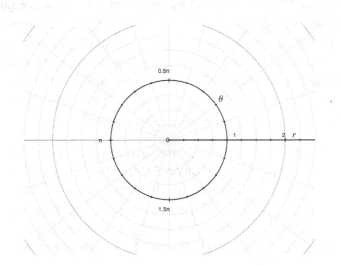

To explore polar coordinates, use `axes.scad` twice, with `piece =` `"polargrid"` (creates the horizontal, r-θ plane), and then with `piece =` `"zgrid"` (creates the vertical, r-z plane) , seen in Figure 4-8. This piece has a grid of radial and concentric lines. Align it so that one straight side is closest to you, and one is on the left. In polar coordinates, we count out the number of radial marks from the origin (the rod sticking up) to find the value of the r coordinate.

FIGURE 4-8

The polar coordinate $(5, 45°)$ on the r-θ plane

We count radial lines starting at the zero line closest to us in Figure 4-8 to find the angle, usually called θ. The long radial lines on this model are in 15° increments, and there are shorter 5° lines and 1° notches along the edge. If we count five concentric lines from the center and three radial lines from the flat side nearest us, that is $r = 5$ and $\theta = 15° * 3 = 45°$. The size and shape of the area of each box on the resulting grid depends on its radial distance from the center of the coordinate system.

Converting from Cartesian to polar and vice versa uses a little basic trigonometry (Figure 4-9). Start off on a Cartesian plane, and construct a line of radius r at angle θ from the x axis. Then, we can see that $y = r \sin (\theta)$ and $x = r \cos (\theta)$. Table 4-1 summarizes how to convert from a point in Cartesian coordinates and get polar ones, or the other way around. In three dimensions, the vertical coordinate (z) is the same in 3D Cartesian and cylindrical coordinates.

FIGURE 4-9

Cartesian vs. polar coordinates

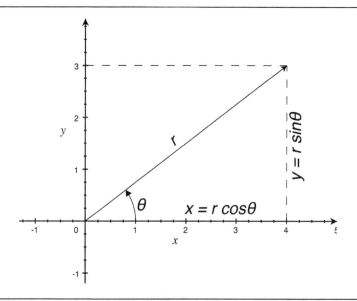

Table 4-1. Converting from Cartesian to polar coordinates

Start with	Use equations	To get
Cartesian (x, y)	$r = \sqrt{x^2 + y^2}$ $\theta = \tan^{-1} \left(\frac{y}{x} \right)$	Polar (r, θ)
Polar (r, θ)	$x = r \cos (\theta)$ $y = r \sin (\theta)$	Cartesian (x, y)

What types of problems might most naturally be solved in each of these systems? If you are used to Cartesian coordinates, it can be hard to get used to the interactions in polar coordinates. In Figure 4-10, we have drawn $r = \theta$, which is called an Archimedean spiral. It is named for the Greek mathematician, engineer, and all-around problem solver Archimedes who lived around 2400 years ago. A 3D variant, the *Archimedean screw*, was an early and efficient water screw, among other applications.

In Chapter 5, we use polar coordinates as a base to learn about what happens when sine and cosine are no longer just tied to a right triangle, but take on definitions for all angles. At that point we will also see what the graphs of sine and cosine look like in polar coordinates.

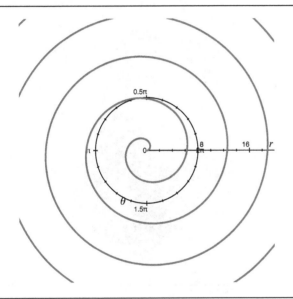

FIGURE 4-10

Polar graph of $r = \theta$ (Archimedean spiral)

Most of the time, we would use these other coordinate systems to work with shapes that are naturally simple in one or the other. A circle is just $r = constant$ in polar coordinates, but a square is messy. You will find that many technical fields have their own weird units and conventions that happen to work out well for what they do. Sometimes whoever pioneered a field created some conventions that are awkward now, but that are so ingrained that they are hard to change. Those of us who cruise around sticking our noses into lots of fields have to be careful to check what symbols mean on the way in the door!

Cylindrical Coordinates

In three dimensions, we can extend polar coordinates into *cylindrical coordinates*, which adds a z, or height, coordinate, just like the Cartesian z axis. The three coordinates of a point then are (r, θ, z). Use `axes.scad` to 3D print two models: the polar grid piece (`piece = "polargrid"`) and the vertical grid piece (`piece = "zgrid"`).

If we have polar coordinates in the plane of our table, we can rise straight above it with a z axis, like we did in Cartesian coordinates. This is called *cylindrical coordinates*, since lines of constant r, θ, and z collectively create a cylinder (Figure 4-11).

FIGURE 4-11

Cylindrical coordinates

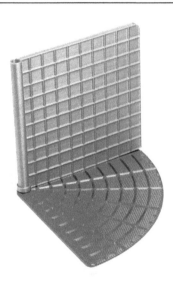

You can try the same exercise as we did for Cartesian coordinates to mark a place on the r-θ plane and then find a height z above it. Take the wedge-shaped piece you used for the 2D polar coordinates, and partner it with the vertical piece you used in 3D Cartesian coordinates.

Use the same vertical grid as before to now find a point in terms of (r, θ, z). For example, to find the point $(r = 5, \theta = 45°, z = 3)$ we would count 5 radial marks out from the origin, 45° counterclockwise from the line closest to you, and three marks up the vertical grid (Figure 4-12). Note that we can also use the vertical piece to show the plane of a constant value of theta.

FIGURE 4-12

Finding a point in cylindrical coordinates

Note that rotation about the z axis is measured with the θ coordinate. However, rotation about axes other than z is tricky to write down. In Chapter 5 we will explore what happens when we move around past $\theta = 90°$.

The distance to the marked point on Figure 4-10 ($r = 5$, $\theta = 45°$, $z = 3$) from the origin is $\sqrt{r^2 + z^2}$, by the Pythagorean Theorem. Calculate what the distance should be, then use a piece of string to measure from the origin to the relevant point in space. Measure the length of the string and see how close you are.

If you do not have access to a 3D printer, another option is to find printable graph paper online and once again add a vertical piece of graph paper taped to a box for the z axis. You also need to be sure that the vertical graph paper has a line that is aligned to the center of the polar graph paper, and that the scale of the r axis on the polar graph matches the scale of the Cartesian graph paper.

Spherical Coordinates

Alternatively, we can use *spherical coordinates*, which add another angle to the mix. It is usually represented by the Greek letter phi, ϕ, pronounced either "fee" or "fie" (rhyming with "tea" or "tie" respectively), with passionate advocates to be found for either pronunciation. The angle

ϕ is measured in the plane at right angles to the (r, θ) plane, and the coordinates of a point in spherical coordinates are given in terms of (r, θ, ϕ).

To make yourself a model to visualize spherical coordinates, use `axes.scad` with `piece = "polargrid"` for the base, and `piece = "altgrid"` for the vertical (altitude) dimension (Figure 4-13). See how small the steps get near the poles of the sphere and how variable the sizes and shapes of the elements are as you move around the coordinate volume.

FIGURE 4-13

Spherical coordinates

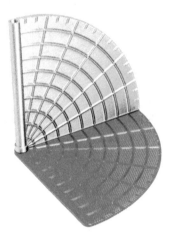

The commonest example of spherical coordinates is the system of latitude and longitude plus altitude on Earth, with altitude being the radius referenced to a mean sea level, and centered on the center of the Earth. That is why latitude and longitude are stated as angles: Pasadena, California, is 34.15° north of the equator (the zero line for ϕ) and 118.1° west of a line that runs through Greenwich in England. The zero line for latitude and longitude are typically given in the order (ϕ, θ) for historical reasons.

If you look at a flattened-out projection of these lines at the pole, you get, not surprisingly, what look like polar coordinates (but which are not stated that way in terrestrial navigation, since the poles are a special case that most of us never encounter). Astronomers have a similar system on the sky, called right ascension and declination, with its own conventions. Sometimes θ is called the azimuth angle, and ϕ the polar angle if measured with the zero line starting at the pole, rather than at the equator.

Different disciplines have different conventions for where to put $\phi = 0°$. Systems using spherical coordinates also need to (somewhat arbitrarily) choose a direction to be $\theta = 0°$ — like the 0° longitude being the line that goes through Greenwich — but that only affects how the coordinates map to physical locations.

The conversion to spherical coordinates from other systems is messy, and requires keeping careful track of signs and a knowledge of which convention to use for the position of $\phi = 0°$. The discussion in Chapter 5 on signs of trigonometric functions will put us on a firmer footing to do those calculations if need be later on. We will use a special case of spherical coordinates in Chapter 12 to describe a robot's movement. The details and all the ifs, ands, and buts are a bit beyond the scope of this book, but formulas for various conventions are available online. The "Spherical coordinate system" article in Wikipedia is a good place to start.

Making Curves and Surfaces

Visualizing surfaces in 3D can be challenging. In this section, we wind up our coordinate system discussion with a little more analytic geometry. We show how to use the coordinate system models to get some intuition about what some relationships among variables look like.

To start, how do you know if an equation represents a point, a curve, or a surface? In any number of dimensions, you need fixed values in all of them to define a single point.

In two dimensions, if only one of your variables is free to vary over all values, the result is a *curve*. This assumes that the other variable is either fixed, or a function of the other variable. For example, $x = 3$ for any value of y, or $x = y$.

In three dimensions, if there are two variables changing over a range of values with the third one dependent on the other two, that would in general define a *surface*, say $z(x, y)$. Sometimes we can have a curve in 3D dimensional space, too (often called a *path*, particularly when there is a defined direction to move along it). These are distinctions we will not worry about in this book.

Rather than trying to develop an OpenSCAD model that would create any arbitrary curve or surface to fit on these template coordinate systems, we will develop two examples and discuss in some detail how we did it.

Craft Materials Models

First, in cylindrical coordinates, you can consider just using a piece of aluminum foil trimmed to the appropriate size to show a surface of constant r. You can also try to show a path in 3D space with pipe cleaners anchored in a bit of clay or tape where they cross an axis. This will not be precise, of course, but can help with reasoning about what a curve will look like in one of the coordinate systems we have provided.

If what follows looks daunting, or you have a roomful of students not quite ready for OpenSCAD, you can always consider using some combination of craft materials to create a surface that will fit on these axes once you figure out where the boundaries will be and a few crucial points. You might bend pipe cleaners or wire to think about curves in 3D space, anchored with some clay or tape, or create a wire frame and cover it with some strips of masking tape for a surface. See Figures 4-14 and 4-15 for ideas.

FIGURE 4-14

The surface r = 3 marked with aluminum foil

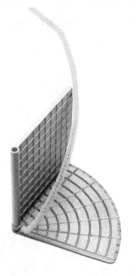

FIGURE 4-15

A 3D curve modeled with some other craft materials

3D Printed Models

We have developed 3D printable models scaled to fit on the cylindrical coordinate system base of the surfaces $z = c\theta$ (where c is a constant) and $r = \theta$. These are generated by the files `helicoid.scad` and `r(theta).scad`, respectively.

So that you have some idea of how to generate a surface if you like, let's talk through how these two one-off surface generation codes work. First, let's explore `helicoid.scad`, which creates the surface $z = c\theta$. A *helicoid* is a three-dimensional shape, the generalized version of an Archimedes screw; see the Wikipedia article "Helicoid" for more. It is also a minimal surface, which means that if we just had a wireframe of the boundary of this shape, a soap bubble would take the shape of the surface we are printing.

The resulting piece is shown attached to the polar grid base in Figure 4-16. (The z axis piece needs to be removed to attach the curve.) We sliced it in PrusaSlicer, using the painted-on supports feature to just have support along one edge and bridge from there. We also painted on a small stretch of support halfway across the lower layers of the surface to add stability. Figure 4-17 shows an early point in the print, with support in the foreground. We also used an attached brim to make the print stick better to the platform. Figure 4-18 shows the completed print, with its support still attached. In Chapter 8 we will explore properties of helicoids a little more and make one out of a soap film.

FIGURE 4-16

The helicoid on the polar coordinate base

FIGURE 4-17

Helicoid model printing

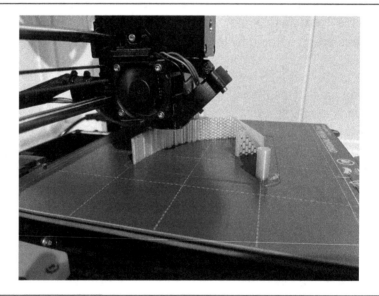

If you are interested in seeing what is under the hood of these models so you can try to make your own, read on. The OpenSCAD model `helicoid.scad` has two basic parts. One creates a tube that will fit onto the polar coordinate baseplate. The other creates a rectangle at $z = 0$ and then sweeps it up (using OpenSCAD's built-in `linear_extrude` function) through an angle of -90°.

We have put in comments (lines starting with `//`) to note the start of each part, and noted the section you might experiment with changing to show the result of sweeping up other lines. Chapter 2 talks about where to find OpenSCAD documentation if you want to get into the details.

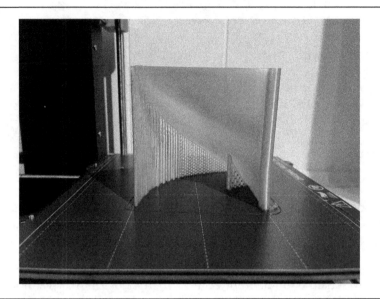

FIGURE 4-18

The completed helicoid print (with support) on the printer

The `tubewall` parameter has been adjusted to be four extrusion widths on the printer we were using (a Prusa Mk3s+). It is essentially the thickness of the vertical part of the print. You might want to tweak that for better adhesion and to minimize travel that might warp the print or rip it off the print bed. Unlike most of our prints in our books, this one is a tad challenging. Here is the entire `helicoid.scad` model. The file is in the book's repository, so you do not need to type it in.

```
// All dimensions are mm
// variables affecting print smoothness
$fs = 0.2;
$fa = 2;
//width of the vertical connector
zthick = 5;
//clearance between connecting parts
clearance = 0.3;
//radial wall thickness of the connector tube
tubewall = 1.6;
linear_extrude(100, twist = -90, steps = 1000) difference() {
  union() {
    translate([-tubewall / 2, 0, 0])
      square([tubewall, 100]);
    circle(zthick / 2 + clearance + tubewall);
  }
  circle(zthick / 2 + clearance);
}
```

The model r(theta).scad uses the similar code to create the tube to attach the print to the coordinate system. However, it is creating the curve $r = b\theta$ for an arbitrary value of z, and a scaling constant, b. This is called a *spiral of Archimedes* in two dimensions; see the 2D graph of a differently scaled one (different value of b) in Figure 4-10. This model does not need support to print, but a brim is recommended. Figure 4-19 shows the resulting print attached to the polar coordinate base.

```
// All dimensions are mm
// variables affecting print smoothness
$fs = 0.2;
$fa = 2;
//width of the vertical connector
zthick = 5;
//clearance between connecting parts
clearance = 0.3;
//radial wall thickness of the connector tube
tubewall = 1.6;
step = 1;
//Define the (scaled) function r=theta
function r(theta) = 100 * theta / 90;
linear_extrude(100, convexity = 5) difference() {
  union() {
    for(theta = [0:step:90 - step])
      hull()
        for(theta = [theta, theta + step])
          rotate(theta)
            translate([r(theta), 0, 0])
              circle(r = tubewall / 2);
                circle(zthick / 2 + clearance + tubewall);
  }
  circle(zthick / 2 + clearance);
}
```

If you change these models to print a different surface, you will need to think through fitting within the coordinate system model (which is 100 mm on a side, or radius 100 mm, depending on the coordinates in use). Both models produce a piece that is sized to fit correctly.

FIGURE 4-19

3D printed curve of r = cθ

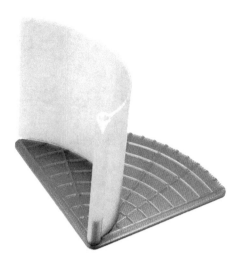

You might try altering one of these models (both in the repository) to try visualizing your own prints, or making the design of curves to fit on the coordinate system models a student project.

Chapter Key Points

This chapter introduced analytic geometry, starting with definitions of Cartesian, polar, cylindrical, and spherical coordinate systems. We discovered that curves and surfaces might be easier to express in one coordinate system versus another, and we explored 3D printed models of the coordinate systems. We saw how trigonometric functions tie together the polar and Cartesian coordinate systems. Finally, we tried out developing a few surfaces to visualize on the 3D printed coordinate system models.

Terminology and Symbols

Here are some terms and symbols from the chapter you can look up for more in-depth information.

- Archimedean screw
- Archimedean spiral
- Cartesian coordinates (x, y, z)
- coordinate system
- cylindrical coordinates (r, θ, z)
- helicoid
- polar coordinates (r, θ)
- phi, ϕ
- radius, r
- slope, m
- spherical coordinates (r, θ, ϕ)
- theta, θ
- y intercept, b

5 The Unit Circle

So far, we have explored sine and cosine in the context of triangles. You have probably also seen wavy curves representing them, too. How are they related? This chapter begins to discuss how sine and cosine curves are derived from triangles constructed inside a *unit circle*, which is just a circle of radius equal to 1. As we find out shortly, this also has implications for the trigonometric functions of angles above 90°. When we graph sines and cosines (and other functions) over a range of angles, it results in a curve that repeats itself periodically, called a *sinusoid*. Let's tie all these things together!

3D printed models used in this chapter

See Chapter 2 for directions on where and how to download these models.

- `cylindergraph.scad`
  - This model creates a 3D unit circle and sinusoid graph.

Other materials needed

- Two sheets of parchment or wax paper, each about 15 inches square
- About 8 ounces Play-Doh or equivalent clay (e.g., two 4-ounce cans)
- Rolling pin or some other implement to roll out Play-Doh into a flat sheet

Beyond the Right Triangle

What happens if you are trying to take the sine or cosine of an angle larger than 90° (or of negative angles)? Those angles can no longer be part of a right triangle, since the two angles that are not the right angle need to add up to 90°. Not surprisingly, the definition gets a little more complicated.

In the last chapter, we learned about the Cartesian coordinate system and how to convert points in it to polar coordinates. A bit of a mashup of polar and Cartesian coordinates is a useful tool to think about what happens when we talk about the sine, cosine, or tangent of an angle over 90°.

Imagine that we have cut an arbitrary right triangle out of red construction paper. Next, we cut identical triangles out of other colors of paper and arrange them around a Cartesian axis so that the same angle of each triangle points toward the center of the coordinate system (Figure 5-1). The equivalent sides of our four triangles are always lined up with the x and y axes. To make our calculations simple, we make the length of the hypotenuse of these triangles equal to 1 in some units. One end of each hypotenuse thus lies on a *unit circle*, a circle centered on the origin of a coordinate system with radius equal to 1. These points are labeled A (red triangle), B (green triangle), C (blue triangle), and D (purple triangle) in Figure 5-1.

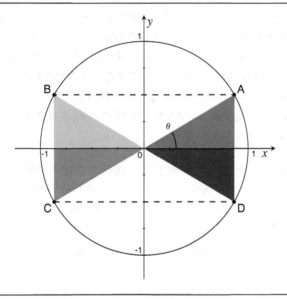

FIGURE 5-1

Finding sine and cosine of equivalent triangles

To find the sine of the angle θ of the red triangle, we need the height of the triangle on the side parallel to the y axis. To get the sine, we would divide this value of y by the hypotenuse (which happens to be equal to 1 here). Similarly, the cosine would be the length in the x direction divided by the hypotenuse. We saw this in Chapter 4's Figure 4-7, when we saw how to convert from Cartesian to polar coordinates and vice versa. As in polar coordinates, the angle θ is measured counterclockwise, starting from the x axis.

Pythagoras would tell us that the hypotenuse would be $\sqrt{x^2 + y^2}$. We already decided, though, that we would have a hypotenuse of length 1, just as we did in our hypotenuse sliding model in Chapter 3. Thus, our formulas for sine, cosine, and tangent of the angle of the red triangle that points toward the center of the coordinate system are

$$\sin (\theta) = \frac{y}{1} = y$$

$$\cos (\theta) = \frac{x}{1} = x$$

$$\tan (\theta) = \frac{y}{x}$$

Angles Between 90° and 180°

What would happen if we flipped Figure 5-1's red triangle over across the y axis? (A mathematician would call this *reflecting* it in that direction.) The result is the green triangle. The angle of this triangle pointing toward the center of the graph is now 180° - θ. Flipping our original triangle over can be thought of as subtracting our angle θ from 180°. In other words, we have to back up θ degrees from the negative x axis, which is 180° from the positive x axis.

The coordinates of the green triangle's highest point (point B) are at the same y value as the corresponding point of the red triangle (point A). Point B is the same distance from the origin along the x axis as A, but in the negative direction. So, we can conclude that

$$\sin (\theta) = \sin (180° - \theta)$$
$$\cos (\theta) = - \cos (180° - \theta)$$
$$\tan (\theta) = - \tan (180° - \theta)$$

Therefore, to figure out the sine, cosine, and tangent of angles between 90° and 180°, we subtract the angle from 180° and adjust the signs accordingly. For example:

$$\sin (120°) = \sin (180° - 120°) = \sin (60°) = 0.866$$
$$\cos (150°) = - \cos (180° - 150°) = - \cos (30°) = - 0.866$$
$$\tan (180°) = - \tan (180° - 180°) = - \tan (0°) = 0$$

Angles Between 180° and 270°

Suppose now we flipped the green triangle in Figure 5-1 over to get the blue one. The blue triangle's x and y sides are pointing in the negative direction. We can think of this as rotating the original triangle to $180° + \theta$. Point C is the same height in the negative y direction as point A is in the positive direction. Similarly, Point C is the same negative x distance as point A is in the positive one. Therefore, we can write

$$\sin(\theta) = -\sin(\theta - 180°)$$
$$\cos(\theta) = -\cos(\theta - 180°)$$
$$\tan(\theta) = \tan(\theta - 180°)$$

Some examples:

$$\sin(240°) = -\sin(240° - 180°) = -\sin(60°) = -0.866$$
$$\cos(180°) = -\cos(180° - 180°) = -\cos(0°) = -1.000$$
$$\tan(250°) = \tan(250° - 180°) = \tan(70°) = 2.748$$

Angles Between 270° and 360° (or -90° and 0°)

Finally, suppose we went back to the original red triangle (where both x and y were positive) and flipped it over (reflected it) to get the purple triangle in Figure 5-1. Now point D has a positive value of x and a negative value of y, both the same magnitude as the coordinates of point A. We can think of this as starting at 360° (or 0°) and subtracting θ.

$$\sin(\theta) = -\sin(360° - \theta)$$
$$\cos(\theta) = \cos(360° - \theta)$$
$$\tan(\theta) = -\tan(360° - \theta)$$

Some examples:

$$\sin(300°) = -\sin(360° - 300°) = -\sin(60°) = -0.866$$
$$\cos(330°) = \cos(360° - 330°) = \cos(30°) = 0.866$$
$$\tan(290°) = -\tan(360° - 290°) = -\tan(70°) = -2.748$$

We can also equivalently think of the purple triangle's angle that touches the center point as being 0° - θ. Negative angles move around clockwise.

$$\sin(\theta) = -\sin(0° - \theta)$$
$$\cos(\theta) = \cos(0° - \theta)$$

$$\tan (\theta) = - \tan (0° - \theta)$$

Finally, if we kept going and went around 360°, we would come right back to where we started. Adding 360° to any angle gives you all the same values for sine, cosine, tangent, and all the other ratios. This means that it is very handy to use these functions for things that are going around in circles, or that repeat somehow.

Principal Value

In Chapter 3, we learned about inverse sine, cosine, and tangent. There, we treated these quantities as having a single value, an angle 90° or less. Now, given the previous section, we know that there are infinitely many angles that have the same values of sine, cosine, etc. since these functions repeat themselves (the official term is that they are *periodic*).

Calculators return the *principal value* of angles. For arcsine and arctangent, this ranges from -90° to 90°. For arccosine, the range is 0° to 180°. That way, there is no ambiguity about what the calculator is returning. Check the physical situation or any drawing you have to see if this is right. Also, be sure you know whether you are working in degrees or radians.

That means though that if you take the sine of an angle greater than 90°, then find the arcsine, you will get the principal value back. For example, sin(120°) = 0.866. However, asin(0.866) = 60°, which is the principal value. Table 5-1 summarizes what the sign should be in each quadrant.

Back in Chapter 3 we noted that the principal value of $\sqrt{4} = 2$. The term applies here similarly, in that if you start off with -2 times -2 you get 4, but a calculator will just give you the result 2 as a default, even though there is a second, valid value for $\sqrt{4}$.

Table 5-1. Sign of trigonometric functions

	0° to 90°	90° to 180°	180° to 270°	270° to 0° (or -90° to 0°)
Sine	+	+	-	-
Cosine	+	-	-	+
Tangent	+	-	+	-

When discussing these functions out loud, it is easy to confuse the words "sign" and "sine." Mathematicians often spell them out (s-i-g-n or s-i-n-e) if an explanation is getting ambiguous. It can feel like trying to talk about a birthday present in front of a two-year-old who is not able to spell yet, but it can help avoid confusion.

The Unit Circle

As we learned earlier in the chapter, a circle centered on the middle of our coordinate system with a radius equal to 1 is called a *unit circle*. If we were to graph $\sin(\theta)$ as a function of θ, particularly for θ ranging up to several times 360°, what would the graph look like?

To do some experiments, we will use a 3D printed model (Figure 5-2), created with the model `cylindergraph.scad`. This model draws a unit circle, and x and y axes on that circle. It also puts tick marks along the margins for regular increments of θ (defaulted to 15°). The model also has, in the vertical direction, a plot of the value of a trigonometric function as you go around the circle. By default, that function is `f(theta) = sin(x)`.

FIGURE 5-2

Unit circle made by `cylindergraph.scad`

Figure 5-3 shows the curve $\sin(\theta)$, which has a maximum value of 1 when $\theta = 90°$ and is -1 when $\theta = 270°$. As we know from our definitions earlier in the chapter, $\sin(0°) = 0$, and $\sin(180°) = 0$. If you added 360° to any value of θ, $\sin(\theta)$ would just be the same value as $\sin(\theta + 360°)$.

Imagine that we were drawing a sine curve as θ went from 0° to 359.999°. If we kept going, we would then start tracing over it again. This is why these are also called periodic functions. As we sweep around the circle varying the angle θ, the values of sine, cosine, and other functions repeat each time we go around a full circle. There are a lot of animations of how you can take all the values of sine and cosine and create a repeating curve that looks like a wave. Try searching online for videos with the keywords "animation unit circle". Since this is *Make: Trigonometry*, though, we will do a little bit of animation with a 3D print and some Play-Doh.

FIGURE 5-3

The graph of sin(θ)

Making 3D Printed Models

To make the 3D printed model in Figures 5-2 and 5-3, open `cylindergraph.scad`, as described in Chapter 2. This model does not use the Customizer in OpenSCAD because we want to change equations. Parameters and equations we change during this chapter are

- The function to plot, `f(theta)`. Examples:
  - `f(theta) = sin(x);`
  - `f(theta) = cos(x);`
- `r`, the radius of the model (mm)
- `2d`, which tells us whether we are making 3D (these models) or 2D (later in the chapter).
  - `2d = false;` will give you 3D.
  - `2d = true;` will give you 2D (a paper print).

To make the sinusoidal function that you want, change the line

```
f(theta) = sin(x);
```

to whatever function you want to plot around the circle. For instance, the graph of

```
f(theta) = sin(2 * x);
```

gives us the model in Figure 5-4. In this case, the curve goes through the sine wave pattern *twice*.

Note that in computer code we have to use an asterisk to show multiplication. The model also has a parameter, `r`, for the radius of the unit circle, if you want to print this out smaller or bigger. It is in millimeters.

FIGURE 5-4

The graph of sin(2θ)

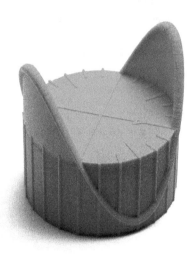

This model is doing a lot of calculation and it can take quite a while —
possibly hours — to render in this version of OpenSCAD. Just start it and
let it do its thing. It creates a string of many small spheres to make the
curve, and generating those is compute intensive. As of this writing, the
current official release of OpenSCAD was last updated in January 2021.
The development version has several experimental features that speed
this up significantly.

Sine Versus Cosine

What about other trig functions? If we change the function in
`cylindergraph.scad` to

```
f(theta) = cos(theta);
```

the result is the yellow print shown on the right in Figure 5-5. The print on
the left is the same sin(θ) print as in Figure 5-3. But wait a minute: Are
they the same, just rotated 90°?

FIGURE 5-5

The sine (left) and cosine (right) prints

Sine and cosine are obviously tied together in a particular triangle. But there is a simple relationship between their respective curves. Suppose we took our red triangle from Figure 5-1 and instead of reflecting it, we rotated it 90° to get the yellow triangle in Figure 5-6. Point A has moved to point E, but all the dimensions and angles of the triangle stay the same. However, now the side of the red triangle that lies along the x direction is in the y direction for the yellow triangle.

The angle of the yellow triangle that is pointing downward in Figure 5-6 is the same as the angle θ in the red triangle. We will call it θ, therefore (a bit different convention than we used in the earlier discussion, but this circumstance is different). Then, for the yellow triangle:

$$\sin(\theta) = \frac{-x}{1}$$

$$\cos(\theta) = \frac{y}{1}$$

FIGURE 5-6

Rotating the red triangle

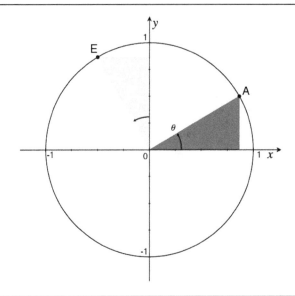

In other words, if we rotate our triangle by 90°, our red triangle's cosine value becomes the negative sine of the yellow one, and the sine value of the red triangle becomes the cosine of the yellow one. This is true in general, and you can see it easily by taking our cosine model and rotating it by 90°, which results in the rotated cosine (yellow) and the original sine (red) to be identical (Figure 5-7).

FIGURE 5-7

The sine (left) and cosine (right) prints

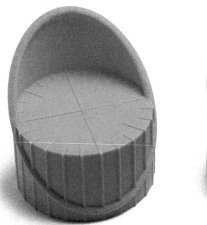

Equivalently, we could have started with the sine and cosine models as shown in Figure 5-5, and turned the sine model 90°. That would have resulted in curves that were the negatives of each other (Figure 5-8) as we would expect from the equations.

FIGURE 5-8

Turning the sine curve

Rolling Out a Sine Curve

Now we will try "unrolling" the curve we have wrapped around the models in the last few sections. First, we take a sheet of parchment baking paper or wax paper and use it to cover a space to work on a flat surface. We used parchment.

Next, we need Play-Doh or equivalent clay. We have found that about 8 ounces (two 4-ounce tubs) is enough. Make it into one big smooth ball, then mush it down a bit so it is easier to roll out (Figure 5-9). Try to avoid having lines or cracks.

Put another piece of parchment on top of the dough to keep it from sticking to the rolling pin. Then, roll the dough out on the parchment in a long thin rectangle, a bit wider than the 3D printed model in Figure 5-8 is tall, and at least its circumference in the other dimension. We used a cylindrical metal water bottle as a rolling pin (Figure 5-10).

FIGURE 5-9

Setup, with model for scale

FIGURE 5-10

Getting the dough flat

Now take the 3D printed unit circle model and roll it along the Play-Doh rectangle. You will get a wavy curve. That is what a graph of a sine wave looks like if you "unroll" it from the unit circle (Figure 5-11).

FIGURE 5-11

The "unrolled" sine wave

One thing is a little tricky, though: the track the model makes when it is used as a roller is the mirror image of the model. Imagine raised or embossed letters on a flat surface. If you stamped the Play-Doh with those letters, they would appear mirrored on the Play-Doh's surface. The same thing happens with these curves.

For a sine curve, you can solve this by starting 180° off from the labeled starting position. That is, you would start rolling with the negative x axis and progress from there (Figure 5-12). More generally, you can create a model for the function of $-\theta$ to create these rolled-out pieces.

Tangent is a bit trickier to graph. Tangent equals the ratio of the two sides of the triangle, the side opposite the angle over the one adjacent. As the angle θ gets closer and closer to 90°, $\tan(\theta)$ approaches infinity since we have a number getting bigger divided by one heading for zero.

FIGURE 5-12

Starting up a sine curve

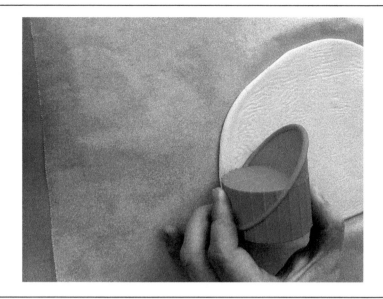

Since there is no way to make a physical model that goes to infinity, we have to cut it off somewhere. We decided, for purposes of this model, to cut off the model when the tangent became plus or minus 3, to keep the 3D printed model a reasonable size. To make a tangent model (shown in Figure 5-13) run `cylindergraph.scad` with

```
function f(theta) = sign(tan(theta)) * min(3, abs(tan(theta)));
```

or if you prefer, equivalently

```
function f(theta) = min(3, max(-3, tan(theta)));
```

FIGURE 5-13

Tangent model

The first part of the equation preserves the sign of the tangent. Now, if we wanted to unroll this one, just offsetting by 180° is not enough, since the curve will be reflected. We can, however, print tan(-θ) (Figure 5-14) to graph tan(θ) correctly (Figure 5-15).

FIGURE 5-14

Models of tan(θ) (left, blue) and tan(-θ) (right, green)

FIGURE 5-15

Rolled out graph of tan(θ) using tan(−θ) model

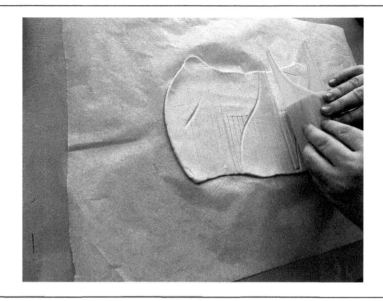

Phase

In the last section, we talked about how to roll our 3D printed models to draw each of our curves. The relative offset of, say, two sine curves that are shifted one from the other, is called a *phase offset*. We have established that sine and cosine have a 90° phase offset from each other. Be careful of which way around the offset is, though! It might seem frivolous to print both sine and cosine models, but it really helps to have both and be able to turn them and think about why they have the alignment that they do. We can write the relationship like this:

$$\cos(\theta) = \sin(\theta + 90°)$$

Phase more generally tells us where we are in a repeating wave. The Greek letter phi, ϕ, is often used for phase. As we go along our curve, our phase will vary, and so it is a function of where we are on the curve, $\phi(\theta)$.

If we are looking at the curve for $\sin(\theta)$, the phase of two points will be equal every 360° since we will have gone "around the circle" once. In the next section we see what happens when our wave repeats more (or less) often than every 360°.

Frequency

What happens to our sine wave if we multiply θ by a constant, like this?

```
f(theta) = sin(2 * theta);
```

That function will go through two full cycles as θ varies from 0° to 360° while the graph of sin(θ) will only go through one cycle. Figure 5-16 shows a model of `f(theta) = sin(2 * theta)` with the radius variable `r = 30` (orange) next to the smaller yellow one, which was created with `r = 15` and `f(theta) = sin(theta)`. The smaller one has half the radius so that it will cover its full cycle in half the distance, to line up with the larger model.

FIGURE 5-16

Models of two different-frequency sine waves

If we roll them out (Figure 5-17) we can see they line up. A wave's *frequency* is how often the wave pattern repeats in some standard interval, like a second, depending on what kind of wave we are measuring. Frequency is typically given in units of 1/time (cycles/second). In the case of our models, it is easiest to think about the waves in terms of cycles per rotation, or maybe cycles per circumference of the model. The circumference determines how far the model will roll per rotation. The yellow model turns twice for one turn of the orange one, because the orange one has twice the circumference. The frequencies are the same when we roll them out because the circumference, and the frequency relative to the circumference, of the orange model are both double those of the yellow one.

FIGURE 5-17

These sine waves rolled out

Amplitude

The default scaling of the model works out to make the height of these waves with different frequencies distinct from each other. That is not any sort of physical constraint in the real world. It is just a default in the model that lets us discern dissimilar waves more clearly.

The maximum height of a wave is called its *amplitude*. In a physical system, amplitude might mean a physical height of a water wave, or might measure some more invisible quantity, like the maximum power of a radio wave. Since the wave goes both positive and negative by this maximum amount, the overall excursion is twice the amplitude.

If the amplitude of a sine wave is five meters, its frequency is two cycles per second, and there is an offset of 45°, the equation of the curve would be written

$$f(\theta) = 5 \sin{(2\theta + 45°)}$$

The difference between the highest crest and lowest point would be ten meters.

This notation can get a little confusing because f is used for "function" as well as for "frequency", and you have to figure it out from context. In practice, people often avoid this by using a capital F for the function, to distinguish it from a lower-case f for frequency, like this:

$$F(\theta) = A \sin{(f\theta + \phi)}$$

In physical applications these waves are often talked about as functions of time (as opposed to as a function of a constantly varying angle of a circle). We can imagine the waves "unrolling" as time goes by as we just did with our models and dough. The frequency of a time-varying wave comes up in many physical applications, like sound waves, where the frequency affects the pitch of the sound, and light waves, where color depends on frequency. We will get into these a lot more in Chapter 8.

Angular Frequency

There is another way of thinking about frequency, called *angular frequency*. It is usually denoted by the Greek letter omega (ω) and involves a factor of 2π since it is measured in radians (Chapter 3). Our wave with an amplitude of 5, a frequency of 2 cycles per second and an offset of 45° (which is $\frac{\pi}{4}$ radians) might then become

$$f(t) = 5 \sin{\left(4\pi t + \frac{\pi}{4}\right)}$$

The 4π comes from the need for our wave to cycle twice as time increments by one second. Different books have different conventions, and the factor of 2π might be implied, burying the conversion to radians and assuming your time variable is scaled appropriately. In that case we would write our function as

$$f(t) = 5 \sin{\left(2t + \frac{\pi}{4}\right)}$$

Physicists notoriously bury all sorts of constants and scaling factors into their equations, so we need to have good habits about keeping units straight when reading physics books. Our *Make: Calculus* book has a discussion of *dimensional analysis*, which is a more formal way of keeping track of this sort of thing.

Period

A related concept is the *period* of a sinusoid, which tells us how long it takes (in distance, time, or whatever units we are using) for the cycle to repeat. The units it is in depends on whether our system is changing over time (measured in seconds), or maybe in physical space (meters). The period equals $\frac{1}{\text{frequency}}$, again with possible factors of 2π depending on our definitions and units. A sinusoid with a period of half a second has a frequency of two cycles per second. In Chapter 8 we talk about the related concept of wavelength.

We do not have to use whole numbers for any of these quantities, and amplitude and frequency can be a number less than one. There are also negative frequencies, but we would need some of the concepts from our calculus book to wade into those. Negative amplitude would just mean a phase shift for simple sine and cosine waves. (Just turn the model in Figure 5-12 to see why). In Chapter 8, we talk about adding and subtracting waves, and how things get more complicated when we no longer have just one pure sine or cosine wave.

Paper Models

The model `cylindergraph.scad` will also print out 2D paper models, designed to be wrapped around something like a toilet paper tube or paper towel roll. When you unroll these flat, they will be correct (not mirrored) since you are not transferring the image, as we did with the Play-Doh. To run these, change the function `f(theta)` to the function you want to graph, and set the variable `2d = true`. A resulting graph is shown in Figure 5-18, rolled up and flat.

FIGURE 5-18

Sample paper graph, rolled up and flat

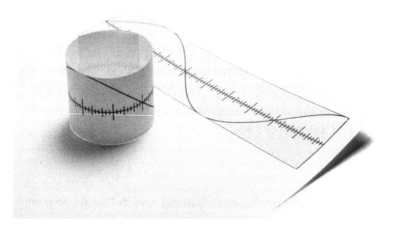

If you do not have a 3D printer, there are a variety of techniques out there for making equivalent models. Cutting off a toilet paper tube at an angle (to make it look like our curve models) is a common suggestion. Try searching online for images and videos with words like "cylinder cut draw sine." (A Russian math site suggests cutting a sausage at an angle, but this seems extravagant!) The cross-section of the curve is a shape called an *ellipse*, and we will learn a lot more about those in Chapter 9.

Graphing Trig Functions

Now that we have seen where these curves come from, there are easier ways to graph them without a rolling pin and dough. For individual values of trig functions, you can use a button on your calculator or paste something into Google for a calculation. However, Google expects angles in radians, not degrees, so you need to convert first, as we described in the sidebar "Degrees, Radians, and Pi" in Chapter 3.

Graphing Sinusoids in Cartesian Coordinates

We have been talking about sine and cosine of an angle, θ. However, that does not mean we always have to plot sine and cosine in a circle. If we "roll out" the functions, as we saw in our example, we get curves in a Cartesian coordinate system. We are then plotting the functions as functions of the x variable. In Figure 5-19 we see that $y = \sin(x)$ (red curve) and $y = \cos(x)$ (green) are gentle curves as the angle increases (horizontal axis), with the signs of each of the ratios varying as shown, repeating the pattern each 360°.

For that reason, we talk about sine and cosine *waves* just as often as we say sine curves; in later chapters we see how these functions are used to model various types of physical waves. The function $y = \tan(x)$ is shown in blue. As we saw earlier, tangent approaches infinity at angles approaching 90° (and 270°, based on the relationships we saw earlier in this chapter, and every 180° thereafter).

FIGURE 5-19

*Sine(red),
cosine(green), and
tangent(blue) for
angles from –360°
to 360°*

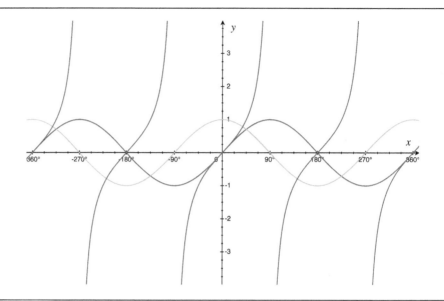

Polar Graphs of Sinusoids

Plotting sines and cosines in polar coordinates gives some interesting results. For example, let's look at the graphs of $r = \cos(\theta)$ (Figure 5-20), $r = \sin(\theta)$ (Figure 5-21), and $r = \cos(2\theta)$ (Figure 5-22). Note that Figures 5-20 and 5-21 draw a complete circle as θ goes from 0 to π , and if we kept going the graph would just repeat endlessly. Finally, if we multiply θ by a constant, like our $r = \cos(2\theta)$ example, the result is a multi-lobed graph which repeats every time θ ranges from 0 to 2π , as in Figure 5-22.

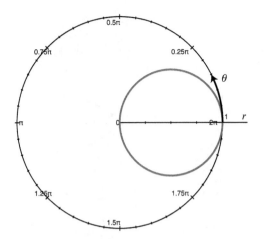

FIGURE 5-20

Polar graph of r = cos(θ)

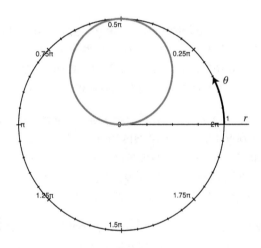

FIGURE 5-21

Polar graph of r = sin(θ)

FIGURE 5-22

Polar graph of r = cos(2θ)

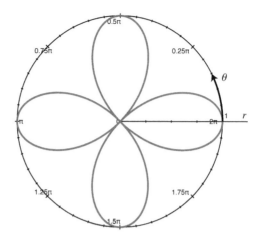

Graphing in Radians

Also, notice that we switched to radians (Chapter 3) in these graphs, instead of degrees. Degrees might be more familiar to start with, but radians often make more sense, even for Cartesian graphs of functions. From here on out, we will mostly use radians.

In general, which coordinate system makes sense depends on the application. For physical applications, one of the most common ways to graph these curves is on a Cartesian coordinate system where the function varies with time. That is, the height of the wave will be the vertical axis, and the horizontal axis will represent time (in radians).

The vertical axis might correspond to physical height (as in water waves) or a more abstract quantity, like how loud a sound is or the strength of an electrical field. The horizontal axis representing time requires a bit more getting used to, as it might seem a little weird to talk about 2π seconds as we measure the period of a wave.

Chapter Key Points

This chapter expanded sine, cosine, and tangent out of the context of right triangles into functions of any angle, including those not between 0° and 90°. We did that by introducing the concept of the unit circle, and understood how the curves for sine and cosine are related to one another. We discovered the concepts of phase, amplitude, and frequency, getting ready for the discussion in later chapters of more sophisticated applications of these curves in analyzing waves and other phenomena.

Terminology and Symbols

Here are some terms and symbols from the chapter you can look up for more in-depth information:

- amplitude
- angular frequency, symbol typically Greek letter omega (ω)
- circular functions
- dimensional analysis
- frequency
- period
- periodic functions
- phase, symbol typically Greek letter phi (ϕ)
- phase offset
- principal value
- sinusoid
- unit circle

6 Trig Identities to Logarithms

3D printed models used in this chapter

See Chapter 2 for directions on where and how to download these models:

- `cylindergraph.scad`
  - This model creates a 3D unit circle and sinusoid graph.
- `rootfinder.scad`
  - This model creates logarithmic scales to simulate a simple slide rule

Optional materials

You may find these items useful.

- Logarithmic-linear graph paper (purchase, or print on paper from online templates)
- Slide rule (we will 3D print one or print a simple one on paper)
- Paper and paper printer

Trigonometry is often taught as part of high school algebra (typically called "Algebra II") but the tie between the two can be obscure. Some of the relationships we saw to this point in the book were known by the ancient Greeks. However, by the 1500s and 1600s, people needed more tools since they were navigating their way across oceans and figuring out the movement of planets. Those tasks required accurate, and extensive, trigonometry calculations. In this hand-calculation era, there was an eager market for faster ways to work with sine, cosine, and tangent.

This chapter is a side trip, introducing some of these handy tools to clean up algebra messes and make calculation easier. Some can cut down on computational loads even today. Others are useful to give us historical insights to build more intuition.

We also introduce a few special cases that we should recognize, if we are lucky enough to be in such a mathematically tidy situation. Failing that, relationships among trig functions can make our algebra simpler, using *trigonometric identities*. These "trig identities", as they are usually called, replace a trig expression with another that is equal to it, but computationally more convenient for the purposes of the moment.

We describe a few identities that we think are the most useful in this chapter. Big tables of them are available online by searching on "trigonometric identity table." The Wikipedia article, "Proofs of trigonometric identities," shows how to prove select identities. If you want an epically comprehensive list (and a lot more math reference material), check out Zwillinger's 2018 book, *CRC Standard Mathematical Tables and Formulas (33rd edition)*, referenced at the end of this chapter.

Proving these identities for the most part requires a lot of algebra, or coming up with a way to cleverly arrange a bunch of triangles to get the right answer. If you would like to see derivations, check out the references at the end of the chapter. We will, however, work through the algebra of one example that led to a light-bulb moment in the early 1600s. This rearrangement of trig identities inspired the development of *logarithms*, handy tools to turn multiplication into addition. This was a huge technological advance in the days before calculators, and was the basis for slide rules. It also will let us see how to take different trig relationships and shuffle them around to get something more useful for the problem at hand.

Law of Sines

The first tool is in a category of its own since it underpins many of the others. It can be stated very simply, too, and is called the *law of sines*. It dates to a bit before the year 1000, and is most commonly attributed to the medieval Islamic astronomer and mathematician Abu'l-Wafa in Baghdad, and the slightly later Persian prince and mathematician Abu Nasr Mensur. The ancient Greeks, particularly Ptolemy, skated around discovering this law, but apparently did not quite get there.

Its statement is straightforward: For any given triangle, even when they are not right triangles, the ratio of the length of each side to the sine of the opposite angle is constant. If angle a is the angle opposite side A, and similarly for angles b and c, we get

$$\frac{\sin(a)}{A} = \frac{\sin(b)}{B} = \frac{\sin(c)}{C}$$

Note that this is not the ratio of the angles themselves, but the ratio of their sines, which will generally be different. This ratio is the same for all three side/angle pairs in a given triangle. The proof gets a little messy; if you are interested, do an internet search on the phrase "proof law of sines."

If we invert each of these fractions, these ratios can also be related to the radius, R, of a *circumscribed* circle. This is the smallest circle that completely encloses the triangle with all the points of the triangle being points of the circle (Figure 6-1). Some authors call this version the law of sines instead.

$$\frac{A}{\sin (a)} = \frac{B}{\sin (b)} = \frac{C}{\sin (c)} = 2R$$

FIGURE 6-1

Law of sines geometry

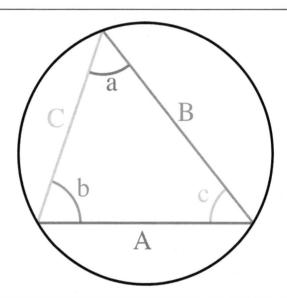

This confirms some things we saw back in Chapter 3: that an isosceles triangle will have two equal sides opposite the two equal angles, and that the side lengths in an equilateral triangle must all be the same for the angles to be the same. It also tells us that a scalene triangle will never have two equal sides.

This relationship underlies a lot of the narrower ones we will meet later in the chapter, and is always a good one to fall back on if we are trying to develop a relationship between angles and sides. As a case in point, let's look at some particularly tidy triangles.

Law of Cosines

There is also a law of cosines, sometimes called the cosine rule. It is a generalization of the Pythagorean Theorem to triangles that are not right triangles. Using the terminology of Figure 6-1, we can write it like this:

$$C^2 = A^2 + B^2 - 2AB \cos(c)$$

Notice that if angle c is 90°, its cosine will be zero. Then the law of cosines just becomes the Pythagorean Theorem. The Wikipedia article, "law of cosines" has several proofs if you would like to see more. If we have a situation where we know two sides of a triangle and the angle between them, it is a convenient way to get the third side. It is the same thing as finding the hypotenuse of a right triangle with the Pythagorean Theorem.

The law holds for all angles of a triangle, so, again referring to the angle labels in Figure 6-1, the following relationships are also true:

$$A^2 = B^2 + C^2 - 2BC \cos(a)$$
$$B^2 = A^2 + C^2 - 2AC \cos(b)$$

In case you are wondering, there is also a law of tangents. It is a little messy and relates the tangents of two angles to their sides in a triangle. It is typically derived from the law of sines.

A final note of caution is in order. If you look up these laws elsewhere, be sure you are reading about these laws for plane (2D) triangles. There is a different law of cosines for angles of a triangle drawn on a sphere. The formula is no longer the same because triangles on a sphere have different properties from those on a flat plane. If you are reading a technical paper, check to see whether the author is talking about geometry on a flat surface or on a sphere if you are trying to follow their trigonometry.

Special Triangles

Some triangles are particularly handy to work with, and being able to recognize them right off the bat can give clues on how to solve a problem. Right triangles with two 45° angles (typically called *45-45-90 triangles*, listing their angles) and *30-60-90 triangles* are often used as examples

(Figure 6-2). They also give us a chance to exercise the law of sines we just learned.

FIGURE 6-2

45-45-90 (left) and 30-60-90 (right) triangles

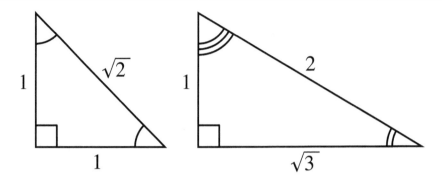

For the 45-45-90 triangle, we know by law of sines that if we have two equal angles, the sides opposite them must be equal. If the sides of the right triangle are both of length 1, then the Pythagorean Theorem tells us the hypotenuse is $\sqrt{2}$.

Using the law of sines, and the definition of sine of an angle as the opposite side divided by hypotenuse, we can read the values we need to check this from Figure 6-2:

$$\frac{\sqrt{2}}{\sin (90°)} = \frac{\sqrt{2}}{1} = \sqrt{2}$$

should (and does) equal

$$\frac{1}{\sin (45°)} = \frac{1}{\frac{1}{\sqrt{2}}} = \sqrt{2}$$

For the 30-60-90 triangle example in Figure 6-2, the shortest side (opposite the 30° angle) is again of length 1. We can use a calculator to find that $\sin (30°) = \frac{1}{2}$. Since sine is opposite/hypotenuse, the hypotenuse must be of length 2. Now from the Pythagorean Theorem (or law of cosines!), we can say that

$$\text{hypotenuse}^2 = \text{shortest\_side}^2 = \text{longer\_side}^2$$

$$2^2 = 1^2 = \text{longer\_side}^2$$

and so

$$4 = 1 = \text{longer\_side}^2$$

Thus, the longer side is of length $\sqrt{3}$, as we illustrate in Figure 6-2. As an additional experiment, let's test out the law of sines with the 30-60-90 triangle.

$$\frac{\sin (30°)}{1} = \frac{\sin (60°)}{\sqrt{3}}$$

Filling in the values for sine,

$$\frac{\left(\frac{1}{2}\right)}{1} = \frac{\left(\frac{\sqrt{3}}{2}\right)}{\sqrt{3}}$$

$$\frac{1}{2} = \frac{1}{2}$$

so this works.

We can confirm that $\sin (30°) = \frac{1}{2}$ in a bit more indirect way by noticing that a 30-60-90 triangle is what we get from cutting an equilateral triangle with sides of length 2 into two equal halves. If we do this with a line from any vertex to the opposite side, we get two of the triangles on the right of Figure 6-2. The side cut in half will be of length 1, and the uncut sides (now the hypotenuses of our right triangles) will be of length 2.

The bottom line is that if we see a problem involving a triangle with a $\sqrt{2}$ in it, we might check whether it is a 45-45-90 triangle. Similarly, values of $\frac{1}{2}$ or $\sqrt{3}$ would make us check for a 30-60-90 triangle. Occasionally a $\sqrt{5}$ might show up, which means you are dealing with a pentagon.

Note that the law of sines does NOT mean we should divide the *angles* divided by the opposite side. For example, in the 30-60-90 triangle case, that would imply $\frac{30°}{1} = \frac{60°}{\sqrt{3}}$ which is not correct. These are simple examples that we can easily verify by other means. The law of sines and law of cosines work for all triangles, and are more useful for analyzing triangles that are not so convenient.

Cofunctions

Next, let's learn a few definitions, in case we are reading a technical book and need to recognize the lingo. First, *complementary angles* are angles that add up to 90°. For instance, since one angle of a right triangle is exactly 90°, the other two angles are complementary to each other (or we say they are *complements*).

If we have a function, like sine of an angle, the complementary angle has a *cofunction*, like cosine, that is the same value. For example, the sine of 30° is equal to the cosine of 60°.

Each trigonometric function has a cofunction. Sine and cosine are cofunctions of each other (hence the "co-") and the cosine of one non-right angle in a given right triangle is the same value as the sine of the other angle. This is true for all the cofunctions.

We have mostly focused on sine, cosine, and tangent so far in this book. The other functions are secant ($\sec(x)$) and its cofunction, cosecant ($\csc(x)$), and tangent's cofunction, cotangent ($\cot(x)$). They are defined, respectively, by these formulas, and are graphed against their more familiar reciprocals in Figure 6-3.

$$\sec(x) = \frac{1}{\cos(x)}$$

$$\csc(x) = \frac{1}{\sin(x)}$$

$$\cot(x) = \frac{1}{\tan(x)}$$

Figure 6-3 has secant on the top (blue) plotted against cosine (red); in the middle we have cosecant (blue) versus sine (red); and finally, cotangent (blue) versus tangent (red) on the bottom. You will note that all these inverses have points where they go off to infinity. We are plotting the x values in radians.

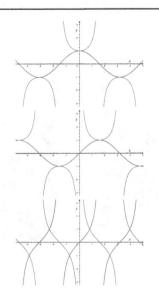

As we noted in Chapter 5, there are other phase relationships among these functions, such as

$$\cos (x) = \sin (x + 90°)$$

Looking carefully at the graph of cot(x) (Figure 6-3) we note that it is the same shape as tan(-x), but with a phase shift of 90°. We would expect this, since, as we saw in Chapter 6, rotating 90° essentially interchanges x and y, thus flipping tangent to cotangent.

$$\cot (x) = \tan (90° - x)$$

Going back to our unit circle model, Figure 6-4 shows the model from Figure 5-14 of tan(-θ) rotated 90° (green) and of cot(θ) (silver) to show the comparison. To create the cot(θ) model, use `cylindergraph.scad` with the function:

```
function f(theta) = min(3, max(-3, 1/tan(theta)));
```

and `r = 30`. We created the tan(-θ) model in Chapter 5, but it is reprinted here in a different filament to show the desired features more clearly.

FIGURE 6-4

Rotate tan(-θ) (left, green) to get cot(θ) (right, silver)

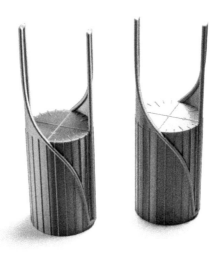

Squared Functions

Now, we will get into a few trig identities that come up often, like those involving squares of sine and cosine, graphed in Figure 6-5. Some useful identities are:

$$\sin^2(x) = \frac{1}{2}(1 - \cos(2x))$$

$$\cos^2(x) = \frac{1}{2}(1 + \cos(2x))$$

Figure 6-5 graphs $\sin^2(x)$ (blue) on the top compared to sin(x) (red). Similarly, on the bottom $\cos^2(x)$ (blue) is graphed against cos(x) (red). We can see that the squared terms are of course always positive. Note that $\sin^2(x)$ is a shorthand way of writing sin(x) times sin(x). If we were talking about the sine of the square of the angle, we would write that as $\sin(x^2)$.

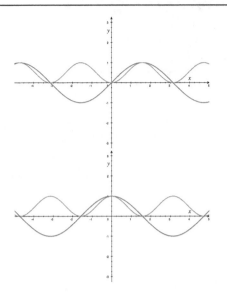

FIGURE 6-5

*Sine (top) and
cosine (bottom) in
red, compared to
their squares in blue*

And speaking of the squares of these functions, one very useful trigonometric identity comes from the Pythagorean Theorem, or if you prefer, from the unit circle (chapter 5):

$$\sin^2(x) + \cos^2(x) = 1$$

Sums of Angles and Double Angles

Sometimes it is helpful to be able to combine angles. What are the trigonometric functions of twice an angle, or the sum of two angles? The formulas for double angles are:

$$\sin(2x) = 2\sin(x)\cos(x)$$

$$\cos(2x) = \cos^2(x) - \sin^2(x)$$

The double angle functions are graphed in Figure 6-6. On the top, we have sin(2x) (blue) compared to sin(x) (red). Similarly, on the bottom we have sin(2x) (blue) versus cos(x) (red). You can see that the phase lines up in both cases when $x = 0$.

FIGURE 6-6

Sine (top) and cosine (bottom) of x in red, compared to the same functions of 2x in blue.

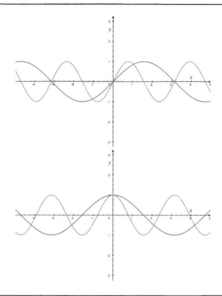

The double-angle formulas are just a special case of the sum of two angles. The formula for this more general case is:

$$\sin (a + b) = \sin (a) \cos (b) + \cos (a) \sin (b)$$

$$\cos (a + b) = \cos (a) \cos (b) - \sin (a) \sin (b)$$

You might think this is all sort of cute but rather obscure. The reality is that these substitutions are power tools you may want to have on your belt if you are going to be doing significant calculations. If you go on to calculus you might find that these relationships can simplify a problem quite a bit. Also, as we see in the rest of the chapter, these identities can have surprising applications.

Prosthaphaeresis

Astronomers, military engineers, and others in the late 1500s and early 1600s were doing a lot of trigonometry calculations. By this time, they had compiled big tables of the functions. There were no computers and calculators, so they had to come up with clever ways to make their lives simpler and to do calculations fast enough to be useful at sea or in the literal heat of battle.

In the 1580s, several people realized that a rearrangement of the formulas in the last section could turn multiplication into addition. This group included François Viète (a lawyer for whom math was a hobby), astronomer Tycho Brahe, mathematician Johann Werner, and several others.

The process that rapidly evolved, *prosthaphaeresis*, was not quite turning lead into gold (another medieval goal). It did however drastically reduce the time needed to do hand calculations, at the expense of a lot of flipping through tables of trig functions. This spinoff of trig identities laid the groundwork for the concept of *logarithms,* which we dig into in the next section.

We need to derive the necessary identity from another one. In this case, we would like to have only multiplication on one side of the equals sign, and only addition on the other so we can convert one into the other. Let's see the types of tricks people use to transmute one identity into another that will be gold for our particular application.

Turning Multiplication into Addition

Start with the sum of angles relationship in the last section:

$$\cos (a + b) = \cos (a) \cos (b) - \sin (a) \sin (b)$$

Then check out the result you get for cos(a - b), by replacing b with - b.

$$\cos (a - b) = \cos (a) \cos (-b) - \sin (a) \sin (-b)$$

We know that cos(-b) = cos(b), while sin(-b) = -sin(b). Taking advantage of that, we get

$$\cos (a - b) = \cos (a) \cos (b) + \sin (a) \sin (b)$$

A common trick in trig identities, if we want to make an inconvenient term go away, is to figure out how to create two versions of the identity and combine them somehow. For example, we might add these two versions together, adding the left sides to each other and the right sides to each other. We can always add something to both sides of an equation. The trick here is that we are adding different, but equal, terms to both sides, instead of the more familiar tricks like adding 3 to both sides, or multiplying both sides by 2.

For example, let's see what happens if we add the equations for cos(a - b) and cos(a + b) together. We do that by adding the left sides of both equations together, and similarly the right sides. First, we start with the two identities:

$$\cos (a - b) = \cos (a) \cos (b) + \sin (a) \sin (b)$$
$$\cos (a + b) = \cos (a) \cos (b) - \sin (a) \sin (b)$$

Then the sum of these two identities becomes:

sum of left sides = sum of right sides

$$\cos (a - b) + \cos (a + b) = \cos (a) \cos (b) + \sin (a) \sin (b) + \cos (a) \cos (b) - \sin (a) \sin (b)$$

The sine terms cancel out, finally giving us this identity:

$$\cos (a - b) + \cos (a + b) = 2 \cos (a) \cos (b)$$

Dividing everything by 2, the final form is:

$$\cos (a) \cos (b) = \frac{\cos (a - b) + \cos (a + b)}{2}$$

If we have sines to multiply, we can get a product of sines by subtracting one identity from the other instead of adding them.

Why is this a good thing? Because, as we noted earlier, multiplication takes a lot more steps than addition when you are doing this by hand. We have a sample calculation in the sidebar. We can see how much work it was to do even minor calculations in those days. Remember that this method was *easier* than multiplying 5- or 6-digit numbers together by hand. For more on the math of this period, check out Uta Merzbach and Carl Boyer's *A History of Mathematics* (3rd edition).

Mercifully, this technique was not the pinnacle of technology for very long. It served as an inspiration for Scottish mathematician John Napier (1550-1617), English mathematician Henry Briggs (1561-1630) and other mathematicians and astronomers. In the 1610s and afterwards, they built on several collective insights to create *logarithms*.

Sample Calculation

Let's try using this technique to multiply 1053 times 345. First, we need to shift the decimal point on our numbers to make them lie between -1 and 1, so we can treat them as cosines. Thus 1053 becomes 0.1053, and we remember we need to move the decimal place of the result by four places when we finish. Likewise, 345 becomes 0.345, moving the decimal point three places (in modern notation — Napier invented that later, as we see in the next section).

The angle whose cosine is 0.1053 is 83.956°. The angle whose cosine is 0.345 is 69.818°. We can now use our formula with a = 83.956° and b = 69.818°. (In the late 1500s, they would have had tables of trigonometric functions to 6 or so significant figures.)

$$\cos{(a - b)} = \cos{(83.956° - 69.818°)} = \cos{(14.138°)} = 0.96971$$

$$\cos{(a + b)} = \cos{(83.956° + 69.818°)} = \cos{(153.77°)}$$

which from Chapter 5 is

$$\cos{(153.77°)} = -\cos{(180° - 153.77°)} = -\cos{(26.230°)} = -0.89703$$

Then we plug these results into the right side of the identity:

$$\frac{\cos{(a - b)} + \cos{(a + b)}}{2} = \frac{0.96971 - 0.89703}{2} = 0.036340$$

Now we move the decimal point 7 places (to make up for how we moved the two initial parts of the product three and four places, respectively) to get the final answer of 363,400.

If we multiply 1053 times 345 using our handy modern calculator, we get 363,285. Our Age of Discovery answer is about 0.3% off, when we carry around angles and cosines to 5 digits.

Logarithms

Multiplication (and, by extension, division) are messy to perform. John Napier learned about prosthaphaeresis from his astronomer friends and was intrigued, but also thought there must be a better way. He started down a path that was completed a few years later by Henry Briggs and others to create *logarithms*.

Logarithms are a way of turning multiplication and division into addition and subtraction, without the nasty step of repeatedly looking up cosines and sines. Suppose we pick some number, like 10, to be our *base*. Then, we write down another number as 10 (our base) raised to some power. For example, 100 is 10^2, 1000 is 10^3 and 0.01 is 1/100, which we will call 10^{-2} (pronounced "ten to the negative second power," or sometimes just "ten to the minus two.") Any number to the zeroth power equals 1.

If we multiply two powers of 10 together, we just add their exponents. Multiplying 10^2 times 10^3 equals 10^5, and multiplying 10^2 by 10^{-2} gives us 10^0. Multiplying 100 times one one-hundredth gives us 1. This is all straightforward enough if we have an integer power of ten.

The tricky bit is what to do with all the numbers that are not integer powers of 10. If we allow fractional powers of 10, then any number can be written as a power of 10. For example, $3 = 10^{0.47712}$ and $2 = 10^{0.30130}$. That means that if we add these exponents, 0.47712 + 0.30130 = 0.77815, we should find that $6 = 10^{0.77815}$ (which it does).

The power to which we are raising 10 is called the *logarithm base 10*, or sometimes just the log or $\log_{10}(x)$. Thus 0.30130 is the log of 2 base 10. Now, how do we know how to find the power of 10 of any number? We look it up in a *log table* (or, in our lush modern days, use the log function on a calculator). To reverse the process, we use the 10^x button on a calculator, or type `10^0.47712` in a search engine. To use log tables for multiplication, we need to use them three times: once to look up logs of each of the two numbers we want to multiply, then an inverse log table to see what power of 10 corresponds to the sum of the logs of our numbers.

But where did these tables come from? In the early 1600s, John Napier used the prosthaphaeresis technique to make his astronomy calculations (relatively) simpler. After doing some of that, he obviously wanted something less nasty. He came up with something very close to our modern logs, but with an awkward-to-use base.

Henry Briggs, a math professor at Oxford, came along and worked with Napier. He discussed the idea of using 10 as the base instead. Napier died in 1617 before the pair had much time together, but Briggs forged on anyway and developed tables of logarithms base 10, publishing in 1624. In Chapter 11, in the section "Hyperbolic Trigonometry," we will learn about logarithms based not on 10, but on a number called e (for "Euler's number"). Logs based on e come up often in physical problems, but we will stick to base 10 in this chapter.

Slide Rules

Logarithms also enabled a powerful analog computer that was indispensable for several centuries, until electronic calculators came along. Edmund Gunter was a friend of Briggs, and created a ruler of sorts that had a log scale. He then used a *divider* (a device with two legs for making measurements) to measure off distances along it to add or divide. This was called a *Gunter rule*. An internet image search will show a variety of them.

Variations on this device evolved over the years, almost immediately with a branch to the *slide rule*, in which two log scales slide along each other to do multiplication, division, and other calculations. William Oughtred was an English mathematician and clergyman, and a friend of Gunter and Briggs. He is credited with creating the first slide rule in about 1622, a circular version.

Slide rules are fundamentally very simple, but, like smartphones today, they started to accrete more functions and scales, sort of equivalent to phone apps. By the 20th century they sported many different scales, as we see on the early-1970s device in Figure 6-7. This one has another set of scales on the reverse. The scales on the two sides are lined up so that the user can flip it over to jump between scales. Improved and specialized versions proliferated for centuries until they were displaced by electronic calculators in the mid to late 1970s. Imagine what an astronomer in 1600 would have given for a calculator with sine and cosine functions! If you have a slide rule gathering dust somewhere, grab it now and follow along. If not, we will show how to make one in a little bit.

FIGURE 6-7

A slide rule set up to figure out 1.5 times 2 (next two photos show closeups of parts of this setup)

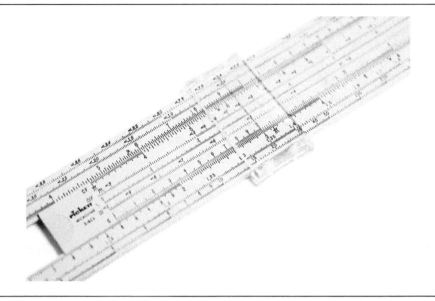

Slide rules can do many things, most fundamentally multiply and divide. Since multiplication is just addition of logs, sliding two logarithmic scales against each other allows a user to just read off sums of logs, and hence the products of numbers. Say we want to multiply 1.5 times 2. We take the "1" on the sliding (middle) part of the rule, and line it up with the first number we are multiplying by (1.5 in this case) as we see in Figure 6-8.

There are two "1"s at the beginning of these scales. Those should be read as 1.0 and 1.1, respectively (since slide rule scales for multiplication start at 1, and not zero). Remember that we are *plotting* the spacing and tick marks corresponding to the log of a number, but we are *labeling* with the number, not its log. Any number to the zeroth power is 1, so the scale starts at 1 (corresponding to a log equal to zero).

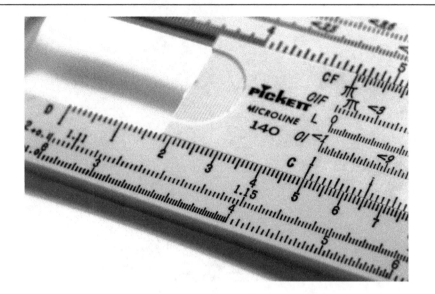

FIGURE 6-8

*Lining up the 1 of
the slider with 1.5 on
the stationary scale*

Then we read along the scale on this same moving part of the rule till we get to the other number we want to multiply (here, 2). Then we read off the answer (3) on the non-moving scale (Figure 6-9). There is a thin red line that moves independently to help with this process.

Just like with prosthaphaeresis, we have to keep track of decimal places separately if we are multiplying two numbers. The logs scale ranges from 1 to 10 (usually also labeled "1") and so the two numbers being multiplied have to be shifted there, too.

We could declare the start of our scale to be 0.1 (whose log is -1), or 0.000001 (a log of -6), but traditionally slide rules are used with the decimal shifted so that the number we are working with is between 1 and 10, and the user keeps track of any decimal shifts. Log of zero is undefined, since there is no power we can raise 10 to and get zero. We can only take the log of a positive number, for the same reason. (For slide rule purposes, we can just take the log of the absolute value, and keep track of signs, like we do for decimal points.)

FIGURE 6-9

Reading off 1.5 times 2 equals 3

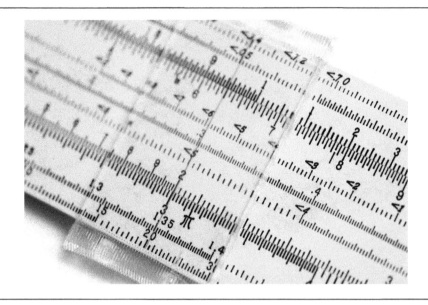

To divide, we do the same thing mechanically with the rule, but read it in a different order. In this case, we would start with the 3 on the stationary scale, and observe that the 2 on the sliding scale is lined up with it. We would go back to the starting point of the sliding scale (Figure 6-7), and see that it is at 1.5. Thus 3 divided by 2 is 1.5.

Powers and Roots

If multiplying in the linear world becomes addition with logs, what happens when you multiply a log? The log of 1000 (which is 10^3) is 3. If we square it, we get 10^3 times 10^3, or 10^6. In general, raising a number to a power is accomplished by multiplying its log by that power. To square a number, multiply its log by 2; to cube it, multiply the log by 3.

A square root is a value that, multiplied by itself, gives you the desired number. For example, the square roots of 4 are 2 and -2, since 2 times 2 and -2 times -2 both give you 4. We will talk about why there are multiple roots later on, in Chapter 10, in the section about the Fundamental Theorem of Algebra. Similarly, a cube root is cubed to get the desired number. The cube root of 27 is 3, for example.

Now, how can we apply this to logs? Taking a square root of a number is the same as raising that number to the one-half power. Thus, dividing a log by 2 gives the log of the square root. Most slide rules have a scale on the stationary part that repeats the log scale on the slider but sized down by a factor of 2. To get a square root on a slide rule, we just line up the original scale and the one that is scaled down by a factor of 2 so the origins are at the same point. Then we can read off the square root. Figure 6-10 shows the process of locating the number 81 on the A scale just above the gap for the slider, and its square root of 9 on the D scale just below the slider gap.

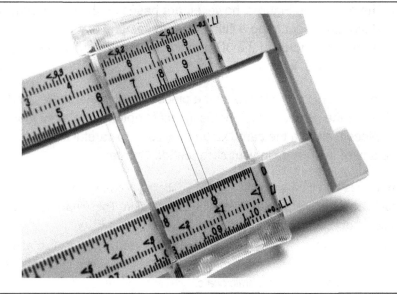

FIGURE 6-10

Reading a square root

Make a Slide Rule

Slide rules might appear to be archaic and completely obsolete. This is true for routine calculation if a calculator or web tool is available. However, they are excellent for getting intuition about logs. If you want to see various slide rules and tutorials on how they work, check out the **online Slide Rule museum** (https://sliderulemuseum.com), particularly their linked "Illustrated Self-Guided Course on How to Use the Slide Rule." This guide includes an interactive virtual slide rule. The site **sliderules.org** has similar resources, as well as a simulated slide rule that you can use in your browser. Wikipedia's "Slide rule" article is a good summary as well.

Having said all that, an actual slide rule is a great hands-on learning device. If you dug one out for the previous section, you might want to continue using it here. Bear in mind, though, that scales on other rules may be labeled differently from the one we show in this chapter.

Otherwise, let's create a simple one with our OpenSCAD model `rootfinder.scad`. This model creates log scales like those of a slide rule one at a time, which we then place next to each other to do calculations.

First, make sure the parameter `2d = false`, which will create a file for 3D printing. To print paper versions instead with a paper printer, see the next section; changing this parameter sets up 3D printing versus paper printing. The next three Figures, however, are based on the paper printing version, which turned out to be a little easier to see and annotate with numbers than were photos of the 3D printed one. The models do not have numerical labels, just the tick marks.

The parameter `scale` is the length of the pieces in millimeters. We used `scale = 150` for the pieces in the photos in this chapter. Then, print four separate pieces, all with the same value of `scale`. Be careful that you do not scale any of the pieces in your slicer either. Create:

- Two pieces with the parameter `root = 1`. Having two of these pieces enables experiments with multiplication and division. These have just one "decade" of logarithmic scale shown.
- One piece with the parameter `root = 2`. This is used in conjunction with one of the `root = 1` pieces to experiment with taking square roots. These have two decades of logarithmic scale.
- One piece with the parameter `root = 3`. This is used in conjunction with one of the `root = 1` pieces to experiment with taking cube roots. These have three decades of logarithmic scale.

Once you have printed the four pieces, take the two `root = 1` pieces. Take one of them and try numbering the major tick marks (starting at 1) and the minor ones as long as you can fit in the number (at the 0.5 mark). You can see how the numbers get closer and closer together as we approach 10 (Figure 6-11). Make sure you have the scale the right way around — the numbers get closer together as we go from 1 to 10. This should help you internalize that the scale runs from 1 to 10, which is important and easy to miss otherwise since we are so used to axes starting at zero.

FIGURE 6-11

One-cycle log scale

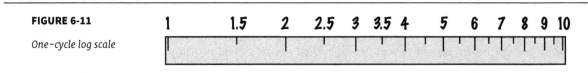

Now, take the second piece that we 3D printed with `root = 1` and line up the "1" (the end) of one piece with the "2" of the other (Figure 6-12). Now, we can read off 2 times any number on the bottom scale on the top scale. You can see how 2 times 2 equals 4, for instance. There are tricks about what to do when your calculation falls off the end of the rule, which are beyond what we want to do here. Check out the websites noted at the beginning of the section if you want to see how it works.

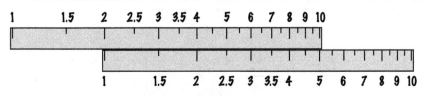

FIGURE 6-12

Multiplying using rules

Now let's try out the `root = 2` and `root = 3` pieces. These have, respectively, two and three *cycles* of log scales (Figure 6-13). In the case of the two-cycle one, the first (leftmost) cycle corresponds to the logs of the numbers 1 through 10, and the second cycle represents 10 through 100. Similarly for the three-cycle one, the first scale (on the left) is 1 through 10, the second is 10 through 100, and the third is 100 through 1000.

Line the rules up carefully, perhaps drawing a line on the left side to help. Label these, as we have in Figure 6-13, and estimate the location of 9 and 16 on the two-cycle scale, and 27 and 64 on the three-cycle scale.

Our two-cycle scale lets us find squares or square roots, depending on where we start, since dividing a log by 2 gives us the square root, and multiplying gives us the square. We can see that 3 on the center scale lines up with 9 on the top scale, since 3 squared is 9. Likewise, 4 on the center rule lines up with 16 on the top rule, which tells us that 4 is the square root of 16.

Similarly, the three-cycle scale gives us cubes or cube roots compared to the one-cycle scale. Moving to the bottom scale, 27 (3 cubed) lines up with 3 on the center scale, and 64 (4 cubed) on the bottom scale lines up with 4 on the center scale. To get really fancy, putting the three-cycle and two-cycle rules together would give you the 3/2 or 2/3 power, again depending on direction.

Play with these models a little to develop more intuition. Users of slide rules had to keep track of decimal points, as we see here. As a side note, it is hard to imagine doing math without the benefit of a decimal point and the concept behind it. However, our current convention has only been around in the West about 400 years, dating back to just before the Mayflower sailed for what would become New England.

FIGURE 6-13

Two-cycle square rule (top), basic one-cycle (center), and three-cycle (bottom)

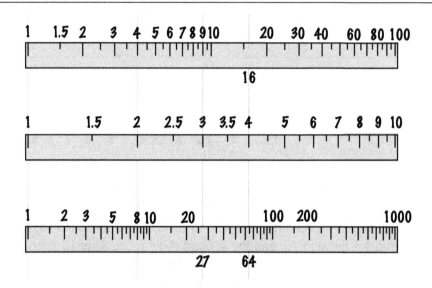

Before that, people worked with a combination of whole numbers and fractions, with various creative and inconsistent means of doing so. (And many of them were still using Roman numerals.) The Arabic world and the Chinese had their conventions too, but hand calculations were painful everywhere back then. Table 6-1 summarizes the correspondence between linear and log operations.

Table 6-1. Linear vs. log operations

Linear operation	Log operation
Multiplication, a * b	Addition, log(a) + log(b)
Division, $\frac{a}{b}$	Subtraction, log(a) - log(b)
Raise to a power, a^b	Multiplication, b * log(a)
Take a square root, $\sqrt{a}$	Divide by 2, $\frac{\log(a)}{2}$

One final historical note, to tie logs and trig functions together even more, is that the availability of logarithms (and tools like early slide rules) allowed people to do more complicated trigonometry than they could by hand. So, a trig identity inspired logs, which then allowed more trigonometry (and ultimately a lot more than that, as we see in the upcoming chapters about trig applications). To read more about that, check out Glen van Brummelen's 2021 book, *The Doctrine of Triangles*.

Paper Slide Rule

If you do not have a 3D printer, `rootfinder.scad` can be used to print out scales on paper instead. As we noted earlier, Figures 6-11, 6-12, and 6-13 are from the version of the model intended to print on paper, versus the 3D print. To create a file intended for paper printing in OpenSCAD, first, set the parameter

```
2d = true;
```

then proceed the same way as we described for 3D prints (that is, make four pieces, two with `root = 1` and one each for `root = 2` and `root = 3`). Instead of exporting each piece to a .stl file, instead export it to a .svg file.

You can open this file in a browser. You might have to use `File > Open` in your browser, or drag it onto a blank page. The file should print from there, or it can be opened in other programs that support .svg, like Adobe Illustrator. As with the 3D prints, be sure that the paper printer is set to the same scaling for all the pieces. Then just proceed from there with the same experiments as we did with the 3D printed parts.

Logs for Estimation

Suppose that we have data about how one quantity depends on another, and we want to see if there is some *power law* correlation — that is, whether two quantities you are measuring are related so that raising one to a power will give you the other. For example, let's say that we get some data for a quantity y varying as a function of some other variable, x, and we graph it as shown in Figure 6-14. We could scratch our heads and try to see the relationship.

FIGURE 6-14

Original graph

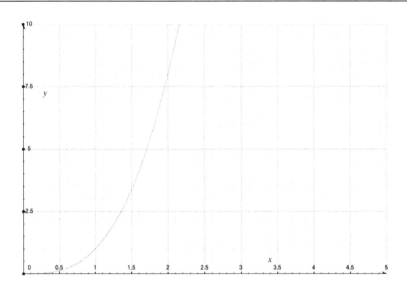

However, if we plot this same function on log-log axes (Figure 6-15) we can read off that the slope of the change in our variable versus the change in time is 3. This means that it might be the case that y varies proportionally to x^3. One always has to return to the principles underlying whatever the data is, but simple graphs like this might give you some intuition about what is going on (or what errors might be masking what you are trying to measure).

Note though that if we had any negative-valued data it most likely would not come over to a log-log plot with most software packages. (Some software might take the absolute value and then plot the log of that, assuming that was what we wanted; but it might not be.) Also, we need to be sure that our log-log graph paper axes give boxes that are the same dimension in x and y, so we can measure off slope correctly. Remember that the slope is not the ratio of the changes in the values labeled, but the ratio of the changes of the plotted logs. Log-log graph paper works too, but be certain to buy or download a version that has the same scales on the two axes.

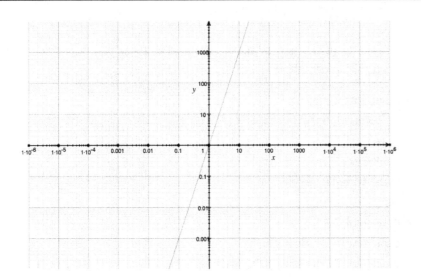

FIGURE 6-15

*Same function on a
log–log plot,
showing a line with
a slope of 3*

Chapter Key Points

This chapter introduced a variety of tools that support trigonometry
calculations. These will help us simplify calculations in later chapters
when we begin to apply trigonometric functions to real problems. We
started off with special cases, and then moved into a brief survey of a few
common trigonometric identities. Finally, we reviewed the historical link
between trig functions and logarithms, and how the resulting functions
are used today.

Terminology and Symbols

Here are some terms and symbols from the chapter you can look up for more in-depth information:

- 30-60-90 triangle
- 45-45-90 triangle
- base
- cofunction
- complementary angle
- cosecant, $\csc(x)$
- cotangent, $\cot(x)$
- cube root
- double-angle formula
- law of cosines
- law of sines
- logarithms, log or $\log_{10}(x)$
- power law
- prosthaphaeresis
- secant, $\sec(x)$
- slide rule
- square root
- trigonometric identities

References

Here are some sources to go into more depth on the topics in this chapter.

The **online Slide Rule Museum site** (https://sliderulemuseum.com) (and its "Illustrated Self-Guided Course on How to Use the Slide Rule") and **sliderules.org** are good resources for further exploration. If you found a slide rule in a drawer and want to use it, searching on the brand name might locate a manual for it, if these websites do not have it already.

The following books are all written at a much more sophisticated level than this one, but will be useful if you are teaching this subject, or if you want to go quite a bit further than we have here.

Merzbach, U. C., and Boyer, C. B., (2011) *A History of Mathematics* (3rd ed.). John Wiley and Sons. This book is a broad resource on the worldwide history of mathematics. It is written assuming that the reader knows mathematical terminology at the college level, but is a very comprehensive guide to use as a reference.

van Brummelen, G. (2021) *The Doctrine of Triangles: A History of Modern Trigonometry.* Princeton University Press. This is a specialized book more focused on just the history of trigonometry. It too is a college-level book aimed more at mathematicians, but we found it interesting for the context of this material. If you are teaching the subject, you might want to dip around in it for some background as you teach various topics. The discussion of how people in the first few decades of the 1600s bootstrapped trigonometry and logarithms to create detailed tables of both by hand will make you hug your computer.

Zwillinger, D. (Ed.). (2018) *CRC Standard Mathematical Tables and Formulas* (33rd ed.). CRC Press. This is a reference book for all things mathematical, with formulas and diagrams for just about any branch of math. As such it might be overkill even for undergraduate students, but it is a good reference for readers who want to branch out, or who need to be sure they are using accurate formulas.

7 Navigation

Materials needed

- Inclinometer project
  - Plastic protractor
  - Drinking straw (large diameter one for drinking shakes is better)
  - Tape
  - Sewing thread
  - Washer or nut (to act as a weight)
  - Yardstick, or some other sturdy straight stick about that length

People have always wanted to get from one location to another reliably. As they explored or migrated farther, navigation needed to get more precise. Trigonometry was at the core of these techniques, and as we saw in the last chapter, was held back in part by the challenges of hand computation. Clever analog computing devices could get around some of this, though, and allow for the rapid analysis needed in a ship at sea or calculating artillery trajectories under battle conditions.

In this chapter, we learn about these analog computing instruments and build the simplest one. We also give pointers to kits to create the more-complex ones, which need parts that might be hard to source one-off (like half-reflective mirrors). We start by revisiting Chapter 3's measurement of the height of a doorframe and garage roof using our new instrument, the *inclinometer*. Next, we see how we can use it to figure out our latitude, with some help from the North Star.

We wind up the chapter with a few high points of navigation history, from traditional sextant measurement to Global Positioning System (GPS) spacecraft and receivers. All the applications in this chapter use the trig functions in the context of the ratios of sides of one triangle. In Chapter 8, we get into applications of the continuous functions such as sine and cosine waves.

Measuring Angles and Distances

In Chapter 3, we saw how to exploit the properties of similar triangles to measure things that are too big to measure directly. That measurement is impractical when more accuracy is needed, and so other instruments have been invented over the centuries to do better. First, we meet the simplest upgrade to our Chapter 3 method, the *inclinometer* (sometimes just called a *clinometer*, or *tilt meter*). There are many variations, but essentially, they provide a way of sighting the angle to a distant point relative to the direction of the force of gravity. The direction of gravity can be found with a weight on a string, a bubble level, or other means.

Inclinometer

In Chapter 3's "Measure Something Big" section, we directly measured the height of a doorframe (a little over 80 inches) and the peak of a garage roof (155 inches). Then we estimated the same measurements by constructing a similar triangle with a known side. As we saw, this is pretty error prone. Can we do any better with a simple inclinometer?

We can make one with a protractor, a large diameter drinking straw, some sewing thread and a weight, like a washer. Take the straw and tape it to the flat side of the protractor. Cut a piece of sewing thread about a foot long. Protractors usually have a hole at the origin point. Tie one end of the thread through that hole, and the other through the washer. Figure 7-1 shows the completed device.

Measuring an angle with it is a little tricky. Calibrate the process by sighting the bottom corner of a doorframe. (We used a different doorframe than in Chapter 3, but one that was also a bit over 80 inches high.) First, we made a little mark on a wall with a piece of tape whose bottom edge was 36 inches above the floor. We lined up the bottom of the tape with the hole in the protractor (also shown in Figure 7-1). Note that the angle of the straw is 90° minus what the protractor is reading, since the protractor markings are relative to an angle at 90° to the way our straw is mounted in Figure 7-1.

FIGURE 7-1

Inclinometer

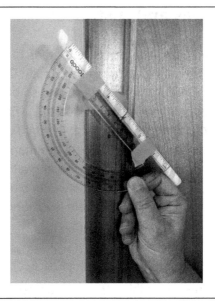

We then sighted the corner of the doorframe and determined we were sighting to it at a 45° angle. We were 45 inches away from the door in the horizontal direction (Figure 7-2). In Figure 7-2, we tried putting a yardstick against the wall and using it as a reference, but the piece of tape visible in Figure 7-1 was easier to deal with.

FIGURE 7-2

Positioning the inclinometer

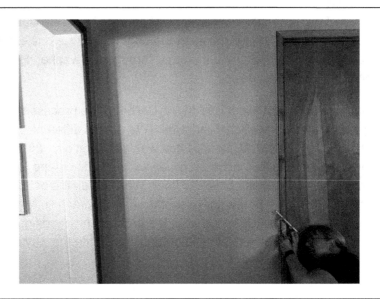

We know that $\tan(45°) = 1$, so the height of the doorframe should equal the distance, since

Tangent of the sighting angle = height of the object / horizontal distance to the object

We have to be careful to keep track of what height we are measuring. In this case, we are measuring the height of the doorframe relative to the height of the hole in the protractor. This hole was 36 inches (a yard) above the floor, so we need to subtract 36 inches from the 80 inch height of the doorframe, or 80 - 36 = 44 inches. As expected, the height of the doorframe above our inclinometer is the same as the distance parallel to the floor to the doorframe, with some measurement error since we are working to whole inches and the doorframe is a bit more than 80 inches high.

This is shown in Figure 7-3, where h1 is the total height of what we are measuring, h2 is the height of the instrument, and h3 is the difference (what we are measuring). In terms of the lengths measured in Figure 7-3,

$$h1 = h2 + \text{distance} * \tan(\theta)$$

Where the angle θ is 90° minus the angle measured by the inclinometer.

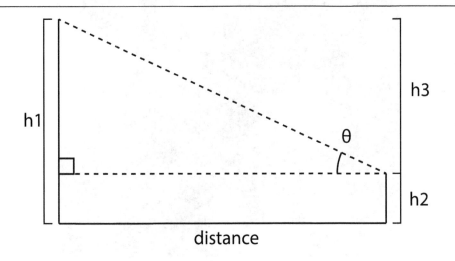

FIGURE 7-3

Diagram of the measurement setup

This all seems a little obvious and contrived for something this simple, but it lets us check our answer before we try measuring something bigger, like the garage roof we measured in Chapter 3.

First, a safety note: Never sight something near the Sun, and never look at the Sun directly! Always arrange your measurements so that the Sun is at your back. If that is not possible, wait until a different time of day. If you want the altitude of the Sun itself, turn your back on the Sun and maneuver the straw until its shadow looks like a ring, with equal-thickness walls all around. The Sun is then shining directly down your straw, but your back is safely to it.

Assuming you can sight your target safely, measure a spot to stand on the ground and mark it (with a piece of tape, or pebble). Write down the distance. Now hold a yardstick or something else of known height to hold your inclinometer steady at this height while sighting. Write down the height of your inclinometer and the angle you measured.

height of object = horizontal distance to object * tan(90° - measured angle) + height of inclinometer

We tried this with the point of the same garage roof as we measured in Chapter 3. This time we stood 191 inches from the garage, measured on the level. The inclinometer read 58°, and we held it on top of a yardstick (Figure 7-4).

FIGURE 7-4

The inclinometer stabilized on a yardstick

The calculated height of the roof peak (in inches, as shown in Figure 7-5) was

$$\text{height} = 191 * \tan(90° - 58°) + 36 = 155$$

This is what we measured back in Chapter 3. The hardest part of this measurement is avoiding swinging the inclinometer when you stop sighting or tilting away from your mark on the ground. It works a lot better if you have two people, one of whom can check that the weight is hanging freely and read off the angle while the other person keeps sighting. If you have access to a tripod and you can invent a way to mount your inclinometer, so much the better.

FIGURE 7-5

Measuring the height of the peak of a roof

To improve on this device, we might include a level as a reference. There are phone apps that act as a virtual level (or give the angle at which the phone is being held), or we can use a *bubble level* as a reference. A bubble level is a bubble in a fluid in a narrow tube. When the tube is level to the ground, the bubble sits in-between two marks. The bubble level reference is at 90° to a gravity reference.

When we consider how else we might make this measurement more accurately, we might end up re-inventing historical instruments. Modern tilt meters or inclinometers are electronic, often made with *microelectromechanical systems* (MEMS) components. These use microscopic devices to detect and measure their position relative to the direction of gravity.

Finding Latitude

Our simple inclinometer can be used to measure *latitude,* which measures how far north or south we are from the equator. (For a lot more about making measurements of latitude and how the Earth orbits the

Sun, see our *Make: Geometry* book's chapter "Geometry, Space, and Time.") Latitude and longitude (north-south distance) create a spherical coordinate system on the surface of the Earth, as we introduced in Chapter 5.

Polaris

Conveniently, sighting a single star will give us our latitude, at least in the Northern Hemisphere. First, we have to find this one star, which is called Polaris, the Pole Star or North Star. It is called that because the North Pole of the Earth more or less points directly at it. That means that if we are sighting through our inclinometer straw at it, our straw is parallel to the Earth's axis and pointing north, as shown in Figure 7-6.

FIGURE 7-6

Geometry of finding latitude from the North Star

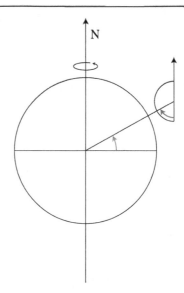

Our string with the weight at the end points toward the center of the Earth, making an angle that is equal to our latitude (blue). This angle is also 90° minus the angle marked on the protractor (red in Figure 7-6). We can see that the angles marked with the red and blue arrows need to add to 90° since they form a right triangle, formed by extending our straw south and extending a line through the equator.

By convention, the equator is latitude 0°, and if we were at the equator the North Star would be almost on the horizon. Our washer would be hanging straight down, reading 90° on the protractor. At the North Pole the straw would be sticking straight up, and the weight would hang along the straw, or 0° on the protractor. In practice it is hard to use this method within plus or minus 15° of the equator, since Polaris is low on the horizon.

Assuming we are far enough north, though, if we were to take a time-exposure image of an entire night, the other stars' apparent motion as the Earth rotates would show up as concentric circles almost centered on Polaris. This means that we can do this measurement any time it is dark enough to see the stars, any time of year. There are some nice visualizations of this on the **Earthsky** (https://earthsky.org) website, in the article about circumpolar stars. We can see why in Figure 7-7's illustration of the position of the North Pole as the Earth goes around the Sun. Polaris would be off to the top left in this illustration, so far away that the North Pole shown in all four orbital positions would always point to Polaris.

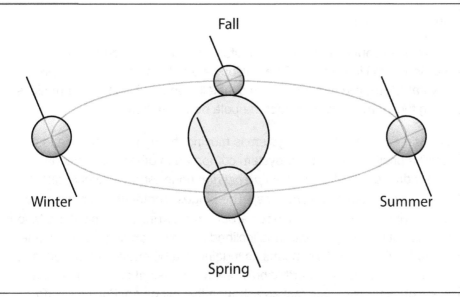

Fall

Winter

Summer

Spring

FIGURE 7-7

The orientation of Earth's axis relative to the stars

Medieval astrolabes (which we explore more in the next section) took advantage of this apparent rotation to calculate time at night. Given the latitude of the observation, the stars would wheel around Polaris over the course of the night. (And the day, too, but invisibly.) Turning the various parts of an astrolabe could indicate what time of night it was by matching up what was visible with a built-in star map, the planisphere. Various aids to making these observations, or observing differences between objects in the sky, evolved over time. This resulted in more applications for astrolabes.

As Earth goes around the Sun each year, the angle of a given latitude to the Sun will change, but the poles point the same places relative to the other stars (Figure 7-7). Polaris would be off to the upper left in Figure 7-7, and we can see that the pole will always point there. Thus, measuring Polaris is independent of time of year, making it an unchanging reference.

In ancient times an astronomer would need to know where Polaris was to get started. Today, we can use a planetarium phone app (or an astronomically savvy friend). We can use the fact that the Big Dipper's handle points to Polaris. Once we have found Polaris, we sight it through the straw and record where the string is marking the protractor. That should be your 90° minus your latitude. For example, at approximately latitude 34° where we live, we would expect the protractor to read 90° - 34° = 56°.

Southern Skies

There is no convenient South Pole star, so this method does not work in the Southern Hemisphere. The Wikipedia article, "Celestial Pole" has several alternative methods using constellations visible in southern skies, using two or more stars to infer the pole between them.

Over time, the whole solar system is moving through the galaxy, and the relative position of the solar system compared to other stars changes slowly due to two effects called *precession* (long-term variations) and *nutation* (shorter-term variations). The biggest contributor to nutation is the gravity of the Moon as it goes around the Earth. The orbit of the Moon is both not perfectly circular and inclined at a bit of an angle to the plane of the Earth's orbit. This means its tugging is a bit erratic. All this adds up, so that the direction of Earth's pole relative to distant stars varies on a scale of centuries, quite fast as astronomical things go. Polaris will no longer be our pole star in a few thousand years.

As a side note, Polaris is now known to be a triple star, made up of three stars that orbit each other. One of the component stars is a *Cepheid variable star*, whose brightness varies in a particular pattern. As it happens, these stars are also helpful for establishing distances since they serve as a *standard candle*, a known brightness that allows distance to the star to be inferred. Thus, it also lets us orient ourselves in the greater galaxy beyond the solar system, too.

Sailing Down the Latitude

Coming up with a latitude number might not seem like a big deal. However, in the era before accurate clocks, it was enough to sail across oceans. Sailors would know the latitude of their destination and would maneuver themselves there. Then they would sail along the latitude line, called *running down a westing* (or *easting*). An astrolabe or other instrument would give approximate time and thus approximate longitude.

Gradually clocks and other technologies allowed measurement of longitude accurately, too. If you want to know more about finding your position based on a few measurements of astronomical objects, check out our *Make: Geometry* book. It discusses how to measure longitude using a clock, some tables of offsets of the Sun's position as a function of time of year, and a measurement of the Sun's shadow. To learn more about the history of longitude in more depth, you might look at Dava Sobel's book, *Longitude*. Finally, should you find yourself in Greenwich, England (near London), be sure to visit the Royal Observatory's museum, which has many amazing timekeeping instruments from the age of sail on display. Some of these are on their website, at **rmg.co.uk**.

Astrolabe

An *astrolabe* is an ancient instrument which is basically an inclinometer optimized to find positions of stars or planets relative to various reference points. In essence, it is an analog calculator (like a specialized slide rule) to sight a known star at night. To help figure out what star was being sighted, usually a sky map (called a *planisphere*) was part of the design. Instead of having a weight, part of the device was aligned to a star or other feature, and typically adjusted for the user's latitude. There was no one specific design, but rather a general evolution to solve various astronomy and navigation problems in the ancient Arab world and Greece, and later in Europe.

The ancient Greeks, notably Apollonius of Perga about 2200 years ago, started with a *dioptra*. This is sort of our inclinometer but mounted on a stable stand. They added a planisphere to get an early astrolabe. These instruments increased in complexity and were often arguably works of art.

For example, the astrolabe in Figure 7-8, from the Smithsonian's collection, was made in Nuremberg by Georg Hartman in 1537. The

Smithsonian notes that there is an inscription inside it that says it also belonged to the astronomer Galileo Galilei. The same instrument is shown in an exploded view in Figure 7-9. Turning the various plates next to each other would allow for various relative measurements. If you would like to learn more about this instrument, the Smithsonian has an entire freely available 1984 piece, *Planispheric Astrolabes from the National Museum of American History,* by Sharon Gibbs with George Saliba. Full download details for it at the end of this chapter.

FIGURE 7-8

Hartman's Planispheric Astrolabe. National Museum of American History, The Smithsonian Institution, Gift of International Business Machines Corporation. CC-0 license (public domain)

FIGURE 7-9

Parts of the astrolabe in Figure 7-8, from the same source

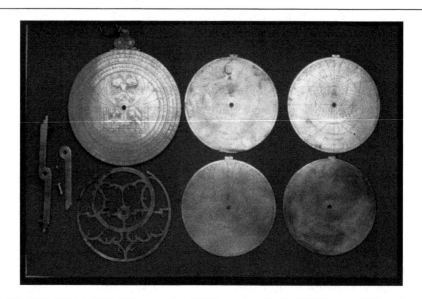

To make your own with a star map requires downloading a template appropriate for your latitude. There is a set of instructions on making your own on the **In-the-sky site** (https://in-the-sky.org/astrolabe/), or you can run the **underlying python scripts** (https://github.com/dcf21/astrolabe) by Dominic Ford. It is based on an astrolabe that was originally designed in 1391!

Ford notes that the device does not work within 15° (plus or minus) of the equator, since Polaris is not high enough above the horizon there. It requires printing out material on paper, with one sheet printed on acetate sheets (overhead transparency film). If you try it, be aware that you need special transparency paper for inkjet printers, or you might need to have that piece printed by someone with a laser printer appropriate for acetate film.

Sextant

The *sextant* was a big improvement over the inclinometer. This device measures angles between the horizon and an astronomical body, like the Sun, or between two objects. The name comes from the Latin word for one-sixth, since a sextant typically is made of a metal 60° wedge. Figure 7-10 shows a sextant created in approximately 1891. Sextants avoid the need for a weight pointing down toward Earth, which is a big improvement over an inclinometer for ships rolling on the sea.

FIGURE 7-10

Plath sextant, c. 1891, NOAA Photo Library (public domain image)

The sextant has two mirrors. One, the *horizon mirror,* is half-reflective (the right or left side of the mirror will be clear). This mirror is lined up with a telescope, and the navigator can look through the clear part to see the horizon. The fully reflective mirror, called the *index mirror,* is mounted on an arm that swings, typically with additional gearing, to allow very tiny adjustments. A celestial object's light reflects off the index mirror and then reflects again off the reflective side of the horizon mirror back to the user.

Figure 7-11 shows the path that light initially takes from the star (yellow arrows) and from the horizon (blue arrows). Light from the two sources comes to the navigator side by side (green arrows), reflected and transmitted respectively, through the halves of the horizon mirror. To use the sextant, the navigator looks through the telescope, and thus through the horizon mirror.

FIGURE 7-11

A sextant's light path

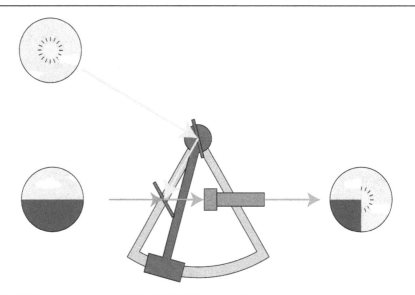

He then adjusts the angle of the moveable arm (red in Figure 7-11) carrying the index mirror until the reflection of the celestial object lines up side-by-side with the horizon in the horizon mirror. Then, the final position of the movable arm when the two are aligned gives the elevation angle of the object in the sky.

There are many variations on these designs, with different ones optimum for certain ambient conditions and tasks. As a side note, in most designs the arm swings through an angle twice that of the elevation angle because of the double reflection. Scales on the sextant are scaled and labeled taking this into account.

In Chapter 8, we explore how light bounces off mirrors and preserves the *angle of incidence* and *angle of reflection,* which is critical for making sextants work. But for now, we can just accept intuitively that light hitting a mirror at an angle reflects off it at an equal angle, relative to a line at right angles to the mirror (Figure 7-11)

Navigational Uses

If a navigator wants to read the elevation angle of the Sun, a thick solar filter will be rotated into place (you can see them at the top of Figure 7-10) so the user is not blinded. If you buy a cheap one to play with, it is best to only use it to look at stars and night, to avoid trusting solar filters that might be intended to be decorative and not functional.

Sextants can be used at night to measure the height of stars, also useful for navigation as we shall see later in the chapter. Sextants are still in routine use by surveyors, usually mounted on tripods. They can also measure the angle between two objects.

The sextant has the advantage over the inclinometer that the user does not have to keep track of a weight or other way of measuring gravity. This has the bonus that it is easier to use on a rolling ship (or, in the early days of flight, on aircraft). As we did for the inclinometer, height of the observer above the horizon line, and other factors, need to be considered for super-accurate measurements.

Historic instruments like the one in Figure 7-10, if used with skill and an accurate-enough timepiece, could measure angles well enough to determine position at sea within a nautical mile. These instruments need to be used in conjunction with good astronomical tables, an accurate timepiece, and navigational charts. The tables and navigational charts of course required trigonometry to create in the first place, too. Modern sailors usually keep a sextant and know how to use one in case electronic means of navigation fail for some reason.

If you want to try making one yourself, a sextant is a more challenging build than the inclinometer. A lot of the "build a sextant" internet projects are really inclinometers. There are various kits on the market to create a true sextant, and pricey historical ones available. There are also very cheap ones meant to be decorative, and probably not safe to look through in daylight. Even if you are a hard-core DIY maker, a kit might make the most sense since it would be challenging to source the optical parts cheaply and to be sure the Sun filters were adequate for the eyepiece scope. (Our tech reviewer, Niles Ritter, did try this; he documented his adventures at **https://www.nilesritter.com/wp/?p=1945.**)

Measuring Height with a Sextant

If we have a sextant, measuring the height of an object, as opposed to just its elevation above the horizon, still requires a direct measurement of the distance from the sextant to the object. Land surveyors use a long tape measure, called a "chain", to determine their distance to an object (just as we did in our inclinometer example), or pace off the distance if that is adequately accurate for the situation. They might also use a measuring wheel device on level ground, if the ground is flat enough.

Laser ranging is a more-accurate way to measure distances. This technique beams a small laser beam pulse at a target, and the instrument measures the time of travel to the target and back. As we will see in Chapter 8, light travels at a finite (very fast) speed — about one foot per nanosecond, which is one billionth of a second. These devices need to be very accurate in their time measurement!

GPS

Using a sextant for navigation has limitations. If it is cloudy or raining, astronomical reference stars are not available. As should be obvious from our experiments with a simple device, it also takes some training and expertise to get good at using these instruments. In recent times, the use of Global Positioning System (GPS) satellites has made it possible to have precise location over much of the Earth's surface.

The GPS system is made up of 24 or more satellites whose orbits crisscross each other around the Earth (Figure 7-12). Each one goes around the Earth twice a day. The constellation is designed so that four (or more) satellites (Figure 7-13) are visible in the sky over most of the Earth. Each satellite has a super-accurate *atomic clock* onboard, which keeps time within three billionths of a second. As a side note, when clocks are this accurate, further corrections need to be made for your phone's velocity relative to the satellite, but that is beyond our concerns here. Atomic clocks measure vibrations of individual atoms. In fact, the international standard for the second is now based on the time it takes a cesium atom to oscillate 9,192,631,770 times.

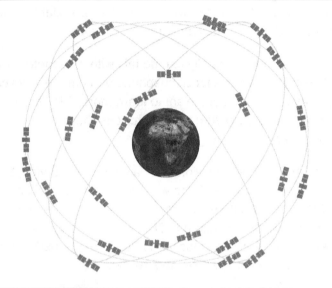

FIGURE 7-12

The Global Positioning System (GPS) constellation in orbit (Public domain image, courtesy US Government, web source (https://gps.gov/multimedia/images/))

FIGURE 7-13

A GPS–III–A satellite (US Government, Public Domain web source (https://gps.gov/multimedia/images/.)

Each satellite broadcasts its position and a very accurate time. Then the user's GPS device on the ground (which has a lesser but still-accurate clock of its own) will compare the position of the satellite and the time at which it received the time from the satellite. If we know the speed of the signal (the speed of light) then the user GPS device knows how far away the spacecraft is. In essence, this draws four or more lines from the satellites to the phone, watch, or whatever device, and we know the orientation and lengths of those lines. The GPS device can use more or less equivalent methods to those we looked at for a sextant (in more dimensions) to find its latitude, longitude, and altitude.

Consumer GPS systems typically can do this with a 10 meter or better accuracy. Making corrections for atmospheric and other effects can get the accuracy even higher. If you want to learn more, GPS manufacturer Garmin has a good **description of the process** (https://www.garmin.com/en-US/aboutgps/).

In the 1980s GPS was not envisioned as a consumer service, but rather as something for military or airliners with big receivers. Ed Tuck, a very determined Southern California entrepreneur and venture capitalist, started Magellan, the first consumer GPS receiver company. The goal was to make cheaper, smaller receivers that would be useful for small aircraft, outdoor recreation people, and the like.

Magellan, founded in 1986, laid the groundwork for ubiquitous GPS applications. Tuck was known for regaling people about the dozens of times the idea was rejected by funding sources, since at the time the perception was that there was no consumer demand for navigation. Of course, time has proven him right, perhaps at the cost of the loss of map-reading skills by anyone under 40. See his essay, *Price, Cost and Staying Alive*, cited at the end of this chapter, for the story behind this modern instance of democratizing navigation.

Chapter Key Points

In this chapter, we looked at applications of trigonometry based on analyzing the angles of one or a few right triangles. This allowed us to understand different types of navigation and surveying problems, and how mathematics and analog computing and observing machines evolved in parallel. In the next chapter we move on to seeing how the sine and cosine continuous functions come into play in optics, sound, and many other fields.

Terminology and Symbols

Here are some terms and symbols from the chapter you can look up for more in-depth information:

- astrolabe
- dioptra
- Global Positioning System (GPS)
- inclinometer (also clinometer, tilt meter)
- latitude
- microelectromechanical systems (MEMS) components
- nutation
- planisphere
- precession
- Polaris (also North Star, Pole Star)
- sextant

References

Online resources for this chapter:

Gibbs, S. and Saliba, G. (1984) Planispheric Astrolabes from the National Museum of American History. *Smithsonian Studies in History and Technology.* 1-231, **10.5479/si.00810258.45.1** (https://doi.org/10.5479/si.00810258.45.1) (It is freely available on a **Smithsonian site** (https://repository.si.edu/handle/10088/2444) as of this writing.) This is a very detailed description of how the astrolabe in Figures 7-8 and 7-9 works, and the history of the particular instrument itself.

Garmin Ltd. (2023) *What is GPS?* **Retrieved 6 January 2023** (https://garmin.com/en-US/aboutgps/). Commercial GPS company with information about its products and a lot of background.

National Coordination Office for Space-Based Positioning, Navigation, and Timing, hosted by NOAA (2022) **Main US government GPS site** (https://gps.gov), with many resources.

Royal Museums Greenwich (https://rmg.co.uk) (2023) Online sampling of collections at the Royal Observatory and Maritime Museums, with a lot of information about the instrumentation of the age of sail.

Tuck, Edward (June 12, 2005) *Price, Cost and Staying Alive*. **Retrieved January 6, 2023** (https://anntuck.com/edtuckessays.html). A philosophy of starting businesses by the founder of Magellan, a consumer GPS company. The discussion of the pricing of GPS units and how they were perceived in the marine market is a classic.

Books:

Sobel, D. (2007) *Longitude*. Bloomsbury USA.

Full sources for the public-domain images in this chapter:

The Smithsonian Institution, National Museum of American History. Hartman's Planispheric Astrolabe. Gift of International Business Machines Corporation, CC-0 license (public domain). **Retrieved 27 Dec 2022** (https://www.si.edu/object/hartmans-planispheric-astrolabe:nmah_214167).

National Oceanic and Atmospheric Administration Library, NOAA Photolib. Plath sextant (1891?). **Retrieved December 28, 2022** (https://photolib.noaa.gov/Collections/Coast-Geodetic-Survey/Other/emodule/1181/eitem/72790).

National Coordination Office for Space-Based Positioning, Navigation, and Timing, hosted by NOAA (2022) GPS images. **Retrieved Jan. 6, 2023** (https://gps.gov/multimedia/images/).

8 Making Waves

3D printed models used in this chapter

See Chapter 2 for directions on where and how to download these models.

- `waves.scad`
    - Create *x*-*y* models of waves in a plane.
- `refraction.scad`
    - Create *x*-*y* models of a few cases of refraction.
- `sinusoidhelicoid.scad`
    - Creates a helicoid

Other materials used in this chapter

- An equilateral triangular glass (or acrylic) prism
- A small flat mirror
- A clear, preferably straight-sided, drinking glass almost full of water
- A pencil that can get wet (will be submerged in water)
- To create a minimum surface in the "Modeling a Helicoid" section:
    - A few feet of 16-gauge uninsulated craft wire
    - Cardboard tube from a roll of toilet paper
    - A tall glass full of soapy water, large enough to hold the wire model

In the last chapter, we saw how trig functions of our humble triangle could navigate us around the world. In this one, we build on our discussions of the functions we derived from the unit circle in Chapter 5 to see a few of the many applications of the continuous trig functions. Our understanding of sound, light, electricity, and magnetism depend on the "rolled out" sine and cosine curves we saw in Chapter 5. These functions are *periodic*, which means they repeat over and over in a curve that often is referred to as a *wave*. Not coincidentally, many waves in nature can be modeled as trig functions.

When we stand on the seashore, watching waves break on the sand, we do not often stop and analyze their behavior. (Surfers and sea captains might be exceptions.) However, similar math underlies ocean waves, the swinging of pendulums, how sound propagates, weather phenomena, and even how electricity and magnetism work.

Unlike most of the math we have seen to this point, these applications of the continuously changing trig functions do not, for the most part, go back to antiquity. Ancient and medieval scholars used trigonometry in the context of computing values associated with a single triangle. They had huge hand-computed tables of trig functions, as we talked about in Chapter 6.

In the 1600s, pieces of our modern view of these curves started to emerge. As he developed calculus, Isaac Newton discovered many principles about sine and cosine waves. Christiaan Huygens analyzed the periodic motion of a pendulum in his 1673 book; our *Make: Calculus* book details pendulum experiments and the math behind them. Finally, Leonhard Euler laid out a lot of what we saw in Chapters 5 and 6 in his 1748 book, *Introductio in Analysin Infinitorum*. If you read Latin or just want to look at diagrams, you can find it **online** (https://archive.org/details/introductioanaly00eule/).

In this chapter, we will lay out a few simple experiments with prisms and lenses and introduce a few 3D printed models. To see where the physics behind these phenomena comes from, we would need to wade into some calculus concepts. However, we can talk about some of the applications without necessarily showing exactly how to compute them.

Sine or cosine waves are not the only game in town. We can invent any type of periodic curve we want. Figure 8-1 shows a few common ones, compared to a sine wave. A *square wave* graphs a quantity that is always one of two values, such as always being equal to one or zero, and changing between the two instantaneously. The voltage of a component that is on or off might be modeled by a square wave.

A *sawtooth* wave ramps up linearly and then drops down instantaneously, over and over. If instead the curve ramps up and then down symmetrically, it is called a *triangle wave*. Square, sawtooth, and triangle waves are often seen in audio synthesis and various other digital electronics applications, where they are easier to generate than true sinusoids.

FIGURE 8-1

*Square (red),
sawtooth (blue),
triangle (green), and
sine (purple) waves*

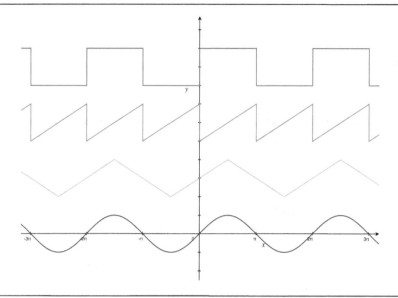

Other physical systems are periodic in more complicated ways. For example, imagine an island with lots of rabbits, and foxes that eat the rabbits. If the rabbit population gets too big, the fox population will get bigger too after a while, since there is more to eat. Then if the overly large fox population over-hunts the rabbits and causes their population to crash, a crash in the foxes' own population won't be far behind. Then the cycle starts over again.

A pair of equations called the *predator-prey* equations, or *Lotka-Volterra* equations (after the people who independently first described them) governs systems that have behavior like this. The equations apply to all kinds of other physical systems where one thing "consumes" another, like chemical reactions. We create models of this system in our *Make: Calculus* book. But they do not involve sine waves.

Adding Waves

Sine waves move up and down in a regular pattern. In this chapter, we talk about "sine" waves to generically mean both sine and cosine, unless we specifically distinguish between them. A more formal generic term for these waves is *sinusoid*.

A lot of their most interesting properties come from what happens when we add multiple waves together. If we drop a rock in a pond and then drop another one right next to it, in some places the ripples will add, and in others they might cancel out. Taking waves and adding them is called *superposition*.

Figure 8-2 shows graphs of, starting at the top, sin(x), sin($2x$), sin($3x$), sin(x) + sin($2x$) and sin(x) + sin($2x$) + sin($3x$) for x running from roughly -3π to 3π. We can see where adding each extra term makes the highest and lowest peaks bigger and shifts where these peaks occur. All the top three waves have the same amplitude (Chapter 5), but different frequencies. We can see that the sin($3x$) cycles up and down three times as fast as sin(x).

Notice that the sums have bigger amplitude, too, as we would expect from adding up curves. They are still periodic, but with a more complicated pattern. We could also change the phase of the curves and then add them up (shift them left or right) or have different amplitudes. In that way, we can create complicated wave patterns.

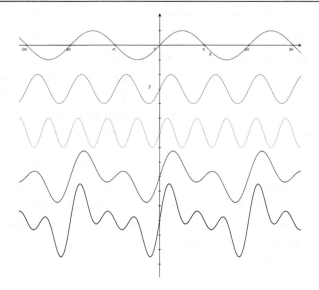

FIGURE 8-2

Adding waves. Top to bottom: sin(x), sin($2x$), sin($3x$), sum top two curves, sum top three curves

Constructive and Destructive Interference

When we add two sine waves, sometimes the peaks will line up, sometimes the valleys will, and sometimes they will partially cancel out. For example, if we added up two sine waves phased exactly π radians (180°) apart, they would completely cancel out. The peaks of one would be over the valleys of the other, and vice versa. Figure 8-3 shows how this works.

FIGURE 8-3

Destructive interference

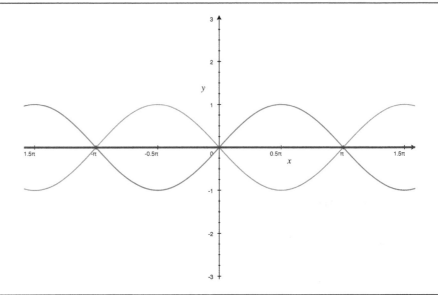

The red and blue curves are sine waves shifted π radians from each other, and the purple line (along $y = 0$) is their sum. We can see that there is exactly as much positive as negative added to the sum at any given point. This is called *destructive interference*, when two waves cancel each other out. Usually of course they do not entirely and exactly cancel, as we can see by looking at the sums in Figure 8-2.

Similarly, *constructive interference* happens when two positive or two negative parts of waves coincide. We see that in Figure 8-2 where there are bigger peaks and valleys in the summed waves than in the originals. Now that we have these ideas down, we can 3D print some models to experiment with interference and think about other physical situations that can be modeled as waves.

Propagation in Space and Time

One thing that is a little tricky to think about is that our waves are propagating in space and in time. Imagine that we were measuring a wave like those in Figure 8-2 at a particular time. We can see a periodic function in a snapshot at any given time, with crests and troughs of waves in space. But as we stand in one place, a wave will be moving past us as a function of time. We glossed over some of these distinctions in Chapter

5 when we just introduced functions like sin(x) or sin(2x), but they become important for many physical applications.

All the wave surface models in this chapter are of necessity a snapshot at a particular time. We would need an animation to see how the waves changed in both space and time. Imagine that we are standing on the side of a highway, and all the drivers strictly follow a rule of each car being 15 feet behind the previous car for every 10 miles per hour they are moving. That is, 15 feet of separation at 10 mph, 30 feet at 20 mph, 45 feet at 30 mph, and so on. We also assume, for the sake of making the math simple, that this is the distance from the front bumper of one car to the front bumper of the next.

This means that the cars will go past us at the same rate (about one car per second), regardless of how fast the cars are moving. Cars moving slowly will pass more slowly, but the following car will be very close behind, while cars passing at a higher speed will be spaced farther apart.

Almost (but not quite) similar ideas apply to our waves. The distance from crest to crest of a wave is called its *wavelength* (often denoted by the Greek letter lambda, λ). The wavelength of a wave is a function of how fast that wave crest is propagating, called the *phase velocity*. It is commonly represented by the letter c by physicists and some mathematicians, and we will use that convention here. The wavelength equals the phase velocity divided by the frequency (often symbolized f). That is

$$\lambda = \frac{c}{f}$$

Or equivalently,

$$f = \frac{c}{\lambda}$$

Frequency, as we saw in Chapter 5, typically has units of 1/*seconds*, and this is often referred to as "cycles per second". One cycle per second is also a unit called the *Hertz* (Hz). It is named after German physicist Heinrich Hertz, who in the 1880s did the first experiments that showed that electricity and magnetism were waves, as predicted theoretically by others. We get into that story in a later section of this chapter. (Note that since Hertz is named after a person, it is not a plural — there is no such thing as a frequency of "one Hert".)

Since velocity has units of length per unit time (for example, meters per second), wavelength has units of length, like meters or millimeters. Frequency can also be in other units in some branches of physics or engineering and the velocity equivalent can mean something else, but we will stick to waves propagating in space as time progresses for now.

In what follows we assume $\frac{c}{\lambda} = 1$ for now, since later we talk about the ratio of the speed of light in air and in a medium, treating the speed of light in air and vacuum as roughly equal to 1 and talking about others relative to it. In most applications, values of frequency, wavelength, and velocity will have a lot of inconvenient zeroes, which cancel each other out. We talk about the real values later, but ratios relative to values in air (that is, setting the ratio of $\frac{c}{\lambda} = 1$) will serve us well for now as we make small physical models of waves.

Waves like these — which oscillate in a direction perpendicular to their direction of motion — are called *transverse* waves. Light and surface waves on the ocean are both transverse waves. We will hear about the other type, *longitudinal* waves, later in the chapter.

Wave Interference Model

The OpenSCAD model `waves.scad` allows us to create models of waves propagating in a plane, instead of just along one axis. Fundamentally, this is not any different from our one-dimensional curve, except that now the height of a wave at any given point is a function of two variables and not one.

The model allows you to either have the waves just on a surface with a flat base (like waves on the ocean surface) or show the wave as a 3D model with both a bottom and top surface. We have done the latter in this chapter, to show the structure more clearly. The parameter `t` (for thickness) controls this and is the thickness in millimeters. We used the value `t = 1.2`, for a 1.2 mm thick wave surface. Setting `t = 0` gives a flat base at `f(x, y) = 0`. To create a model with a flat base for a wave with positive and negative values, we would need to add a constant to all values of `function f(x, y)` so that all values are greater than zero.

The model breaks a 2D surface into points and calculates the height of each point, then creates faces to join adjacent points into a surface. If we try to create a wave that would have its crest and valley in less than the distance between these points, the waves will be distorted, or we might miss them entirely. The shortest wavelength wave (and thus highest frequency wave) we can model with `waves.scad` should be four or so times the variable `res`, which determines how big these chunks are. It is defaulted to 0.5 mm, so waves with features several times that are the finest you should try.

More generally, when looking at physical applications, this sampling limit goes by the name *Nyquist frequency*, after Harry Nyquist, who worked for what became AT&T Bell Labs from 1917 to 1954. The limit is usually stated that one must sample at twice the frequency of the wave being sampled. (In realistic practice, one must sample more than that, usually, for good signal quality.) Nyquist developed a lot of other fundamental ideas in communications theory as well.

The model `waves.scad` requires you to define the `function f(x, y)`, which describes the waves we want to model. We also set a wavelength and amplitude of our waves. We use a feature of OpenSCAD called the Customizer to allow you to pick the appropriate case from a menu. OpenSCAD expects trig functions in degrees, but it is more convenient to think about most of these functions in radians. As a result, we need to put in a conversion factor of $\frac{180}{\pi}$ to change our radians to degrees. Should you look at the actual model in detail, you will see that factor all over the place.

We have printed models of three different waves, all using the same values of the parameters for wavelength and amplitude. The first (Figure 8-4) shows a wave propagating radially from one point. This is called a *point source* (in the Customizer, `point_source`). The equation for this in radians is

$$f(x, y) = 2 \sin\left(\frac{1}{2}\sqrt{(x - 50)^2 + y^2}\right)$$

As we see in Figure 8-4, this is a wave that radiates out from halfway ($x = 50$) along one side of the model. The amplitude is still 2, and frequency is $\frac{1}{2}$. We should mention here that in real life, electromagnetic waves (like light waves) will have this behavior. Water waves, like those resulting from a pebble thrown in a lake, are dispersive, and will spread out and gradually die out. The details, however, are represented by complicated calculus equations and get far from the realm of basic trigonometry!

FIGURE 8-4

Point source wave model

The next (Figure 8-5) shows a wave propagating with straight line waves, called a *plane wave*. The amplitude and frequency remain the same, but now we have a wave just propagating in the y-direction, with a phase such that the sine wave is zero at $y = 50$. The equation we are plotting in 3D in this case (select `plane_wave` in the Customizer) is

$$f(x, y) - 2 \sin\left(\frac{1}{2}(y - 50)\right)$$

FIGURE 8-5

Plane wave model

Finally, the last model is the point source with the plane wave added to and subtracted from it (Figure 8-6). These functions are

$$f(x, y) = 2 \sin\left(\tfrac{1}{2}\sqrt{(x - 50)^2 + y^2}\right) - 2 \sin\left(\tfrac{1}{2}(y - 50)\right) \text{ (subtracted)}$$

and

$$f(x, y) = 2 \sin\left(\tfrac{1}{2}\sqrt{(x - 50)^2 + y^2}\right) + 2 \sin\left(\tfrac{1}{2}(y - 50)\right) \text{ (added)}$$

To plot them, select `wave_diff` or `wave_sum`, respectively, in the Customizer.

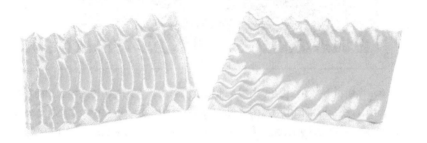

FIGURE 8-6

Point source and plane wave added to (left) and subtracted from (right) each other

These models need to be turned 90° and printed on one edge, since printing them flat would require a huge amount of support and the surface would be a mess. Which edge makes the most sense will depend on which orientation will have the least aggressive overhangs. We used a small raft to help stabilize the print (Figure 8-7).

FIGURE 8-7

Model on a printer

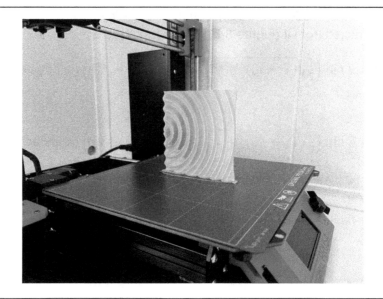

There are many techniques that add together waves to create more-complex curves. For instance, *Fourier transforms* are a magical seeming one that can add up waves of various amplitudes, frequencies, and phases to get almost any other periodic (or not) curve. The details of how that works are beyond what we can do in this book, but the general premise is not all that different from what we have done here. The tricky bit is that a way of doing the process backwards (*inverse Fourier transforms*) can start with a curve resulting from adding these many waves and can give back the simple waves that created it. Actually performing these calculations requires sophisticated calculus techniques.

Electromagnetic Waves

So far, our wave discussion has been a little generic, focused more on the math than on the specifics of an application. Now, let's look at seeing what some of these waves look like in the wild. In this section we look at the basics of light waves, and how they travel through different materials.

Light is a challenging thing to study without precise instruments. The basics of optics were known and applied by Galileo and Newton in the 1600s, but both equipment and mathematics had to develop for a while longer before modern models of how light works came along. Two men — Michael Faraday and James Clerk Maxwell — were major contributors.

Faraday was an English experimental scientist who worked in the first two-thirds of the 1800s. He did not have a formal education, and probably never even learned the math in this book. However, he managed to talk his way into apprenticeships with other scientists. He was famous as a very careful experimenter and made measurements of how a magnetic field could cause electricity to flow. Late in his life, he influenced the Scottish mathematician and physicist James Clerk Maxwell.

Maxwell was able to take Faraday's data and his own observations and create the equations that tie together electricity and magnetism. He was also able to deduce that light waves travel as intersecting electric and magnetic fields at right angles to each other. One basic visualization of a light wave is shown in Figure 8-8. You can imagine that the electric wave is vertical, and the magnetic one horizontal, and that the wave is propagating to the left. Collectively, the electric and magnetic fields make up an *electromagnetic wave*, such as a light or radio wave.

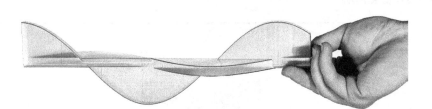

FIGURE 8-8

Electric and magnetic fields

Maxwell's equations are often seen on geeky T-shirts, and they form the basis for the modern understanding of light waves. For a well-written history of the two men, check out the book by Nancy Forbes and Basil Mahon, referenced at the end of this chapter.

Diving into Maxwell's work in depth requires sophisticated calculus concepts. (Figure 8-8 is in fact a calculus model showing other properties of sinusoids, and we do not include it in this book's repository — it is part of our *Make: Calculus* model repository.) However, we create simplified models of some of the basic wave phenomena in the rest of the chapter. You may have heard that sometimes light can be thought of as a wave, and sometimes as a series of particles. In this trigonometry book, we will stick with discussing some basic optical phenomena that will not require us to draw the distinction.

Analyzing light waves makes us think about sine waves (the light wave itself). In some cases, though, we also need a little single-triangle trigonometry to understand phenomena. Two simple effects where this is the case are *reflection* and *refraction*.

Refraction

Now that we have tried modeling waves with no constraints on them propagating forever, and then reflected waves, how about understanding what happens in a prism? Prisms are glass pieces that are usually shaped like an extruded equilateral triangle (Figure 8-9). They are handy for turning light, as we will see.

Before we go any further with this, though, we need to find out how we can model a wave that is at an angle to a surface. The easiest way to do that is to use a *rotated coordinate system*. In the next section we will learn the mechanics of how to do that, but for the moment let's stick to why we need to know.

FIGURE 8-9

A prism

Prisms can turn light through different angles because light moves at different speeds in air and in glass. Even more interesting is that different wavelengths of light will turn differently as they cross into and out of a prism. Ambient white light, like that from a light bulb, is a mixture of different colors of light, which have different wavelengths. The longest-wavelength light visible to the human eye is red light, which is about 660 nanometers. The shortest is violet, at about 480 nanometers. Orange, yellow, green, and blue fall between these extremes, from long to short wavelengths.

There is an entire "electromagnetic spectrum" with wavelengths longer than whole planets all the way down to shorter than visible light. For example, "millimeter waves," have a wavelength between 1 and 10 millimeters, and are what cook your food in a microwave oven. Radio waves can be many meters (or kilometers) long. Some science and engineering disciplines assume a 2π floating around in their wavelength definitions, too, which we have to recognize from context and tradition in various fields.

As we saw earlier in this chapter, we can add up different waves of varying wavelengths to get one more-complicated wave. White light is a summed-up wave like that. A prism can let us split out this mixture of different-color light by wavelength and uncombine this summed-up wave. Let's dig into why this works with a 3D printed model of what the waves are doing.

When we talk about something moving fast, we might say that it moves at the speed of light. In everyday life, light moves so fast that it feels instantaneous. It is exactly 299,792,458 meters per second, or about 186,000 miles per hour. The speed of light in a vacuum is now a fundamental constant in the International System of Units (SI) and is used in turn to define the length of the meter. In vacuum, then, the wavelength of light and its frequency are related like this:

$$\lambda = \frac{c}{f}$$

That means that in a nanosecond (1/100,000,000th of a second) light moves about a third of a meter (or, if you prefer, about a foot). In other media, the speed may be a bit slower. The ratio of the speed of light in a vacuum and in another medium is called the *index of refraction* of the medium. In water, light goes about 1/1.33 (or 75%) as fast as it does in vacuum, so its index of refraction is 1.33. The index of refraction of air is close to vacuum, and of various kinds of glass about 1.5.

Rotated Coordinate Systems

In Chapter 4, we learned how to think about Cartesian coordinates and polar coordinates. What happens, though, if we want to rotate our

Cartesian coordinate systems through some angle to make things more convenient? For example, we want to model waves incident on a surface at some angle. It is easiest to model the surface as aligned to the coordinate system, so that means we need to rotate the modeled wave by the incident angle.

The easiest way to do that is to start off with a point P(x, y) as shown in Figure 8-10. We want to rotate our coordinate axes by θ degrees counterclockwise (in the positive-theta direction). As we saw in Chapter 4, we could change the point P(x, y) into polar coordinates (r, θ) by letting

$$x = r \cos (\theta)$$
$$y = r \sin (\theta)$$

FIGURE 8-10

Coordinate transformation

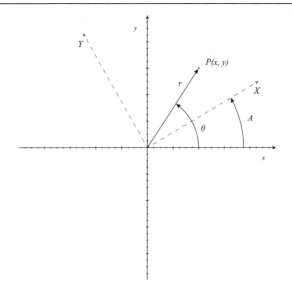

Note that the two coordinate systems have the same origin, so the radius to our point stays the same. Let's call the rotated coordinate axes X, Y, rotated by some angle A from the original coordinate axes. In the new coordinate system, polar coordinates would be

$$X = r \cos (\theta - A)$$
$$Y = r \sin (\theta - A)$$

That is fine, but now what? In Chapter 6, we saw how to deal with the sums and differences of angles. The formulas for the sum of two angles we saw there are

$$\sin (a + b) = \sin (a) \cos (b) + \cos (a) \sin (b)$$
$$\cos (a + b) = \cos (a) \cos (b) - \sin (a) \sin (b)$$

We know that $\sin (-b) = -\sin (b)$ and $\cos (-b) = \cos (b)$. So $\sin (a - b)$ is $\sin (a - b) = \sin (a) \cos (b) - \cos (a) \sin (b)$

Similarly,

$$\cos (a - b) = \cos (a) \cos (b) + \sin (a) \sin (b)$$

(negatives cancel out)

In terms of our θ and A,

$$\sin (\theta - A) = \sin (\theta) \cos (A) - \cos (\theta) \sin (A)$$
$$\cos (\theta - A) = \cos (\theta) \cos (A) + \sin (\theta) \sin (A)$$

Therefore:

$$X = \cos (\theta - A) = \cos (\theta) \cos (A) + \sin (\theta) \sin (A)$$
$$Y = \sin (\theta - A) = \sin (\theta) \cos (A) - \cos (\theta) \sin (A)$$

and we know that our original x and y were equal to r cos(θ) and r sin(θ) respectively, so

$$X = x \cos (A) + y \sin (A)$$
$$Y = y \cos (A) - x \sin (A)$$

The cases we will play with in rotated coordinate systems (reflection and refraction) will stick to waves that are just functions of Y, so we will only need to worry about $Y = y \cos (A) - x \sin (A)$. The angle A here is called the *angle of incidence, refraction,* or *reflection,* depending on what we are doing.

Snell's Law

Now imagine we have a light wave propagating in air that then passed into glass. Its frequency will stay the same, since there are a certain number of waves entering the glass and those do not go away. However, the speed will change, and that means the wavelength has to change. Since the waves are getting closer together, the wave will also turn a bit as it goes into the new medium.

The rule that governs this is called *Snell's Law,* for an early-1600s Dutch astronomer named Willebrord Snellius who derived the law but did not publish it. Descartes, as well as others in both the West and the Arab world, came up with versions of it as well around the same time. Snell's Law is very simple and illustrated in Figure 8-11.

FIGURE 8-11

Angles involved in Snell's Law

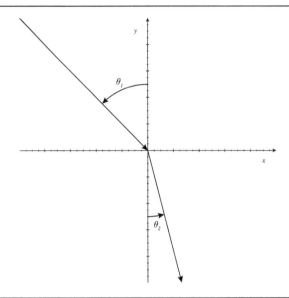

A line perpendicular to the propagating waves is an angle θ_1 relative to a right angle to the surface of our glass, water or other medium. After passing through the interface, a line perpendicular to the new direction of the wave will have the angle θ_2 relative to a right angle to the surface of our medium. Given these definitions, Snell's Law relates them like this:

$$n_1 \sin (\theta_1) = n_2 \sin (\theta_2)$$

Where n_1 is the index of refraction of the first medium, and n_2 that of the second medium. So, for instance if we went from air ($n = 1$) to glass ($n = 1.52$) and we started out with a wave with $\theta_1 = 30°$, then we would find our second angle from rearranging the previous equation to get

$$\sin (\theta_2) = \frac{n_1}{n_2} \sin (\theta_1)$$

and so

$$\theta_2 = \sin^{-1} \left(\frac{1}{1.52} \sin (30°) \right)$$

or about 19.2°.

The $\sin^{-1}$ (...), or arcsin, function is read "angle whose sine is", as we saw in Chapter 3. The wave would turn quite a bit more toward the normal to the surface in glass than it started out in air.

Another tricky bit is that as they turn the waves will get closer together, since they are moving more slowly. (The frequency, as we noted, stays the same.) So, the wavelength will also fall by the ratio of the indices of refraction.

The classic way to see Snell's Law in action is to put a pencil in a glass of water. The pencil appears to bend at the water's surface, since you see the light waves shift (Figure 8-12).

FIGURE 8-12

A pencil appearing to bend at water's surface

Now, let's get really tricky. It turns out that the index of refraction of many materials also depends on the wavelength of the light. Thus, if we have white light (a mix of many different wavelengths) we might be able to use moving into another medium as a handy way to separate out the colors. This is exactly what a prism does.

Let's use `refraction.scad` to first model red light waves going from air into glass, with an incident angle of 45°. We will do some scaling of our waves so that we can sensibly compare the results for different colors. We will scale our wavelengths relative to the wavelength of violet light, which is about 480 nanometers. Red light is about 660 nanometers. We also will introduce a factor of 2 (the absolute scale is arbitrary) to make it print better. Glass typically has an index of refraction of about 1.52. Let's see what we have to add to `refraction.scad` to get this to work.

We are going to print the model for red and violet wavelengths and compare them. Red is the longest wavelength that human eyes can see, and violet is the shortest. Red and violet are the two extreme color bands on a rainbow. We printed these models in PrusaSlicer with the perimeters set to a little thicker than normal (0.6 mm) so that there would not be any open spaces in the models, making them more transparent.

Here is how we set the parameters in `refraction.scad` to model red light. We want to scale our waves relative to violet light at 480 nm, so we are using a wavelength of `660 / 480` for red, and we show a wavelength of `480 / 480` (which of course is just 1) for the violet. The index of refraction can be a little lower for violet, and looking up some typical numbers for glass we use 1.52 for red light and 1.509 for violet.

$angle_{in} = 45°$ (incident angle)

$$angle_{out} = \sin^{-1}\left(\frac{1}{1.52} \sin\left(angle_{in}\right)\right)$$

Propagating wave in air:

$$f(x, y) = 2 \sin\left(\frac{480}{2*660}\right)\left((y - 50) \cos\left(angle_{in}\right) - x \sin\left(angle_{in}\right)\right)$$

Propagating wave in glass:

$$f(x, y) = 2 \sin\left(\frac{480}{2*660}\frac{1.52}{1}\right)\left((y - 50) \cos\left(angle_{out}\right) - x \sin\left(angle_{out}\right)\right)$$

In OpenSCAD, the parameters used are

```
angle_in = 45;
amplitude = 2;
lambda = 2 * 660 / 480;
ray_width = 1;
refractive_index = 1.52;
refract = 1;
```

Next, we change the wavelength and index of refraction to get violet light (case refract_violet in the Customizer).

In air:

$\text{angle}_{in} = 45°$ (incident angle)

$$\text{angle}_{out} = \sin^{-1}\left(\frac{1}{1.509}\sin\left(\text{angle}_{in}\right)\right)$$

$$f(x, y) = 2 \sin\left(\frac{480}{2*480}\right)\left((y - 50)\cos\left(\text{angle}_{in}\right) - x \sin\left(\text{angle}_{in}\right)\right)$$

In glass:

$$f(x, y) = 2 \sin\left(\frac{480}{2*480}\frac{1.509}{1}\right)\left((y - 50)\cos\left(\text{angle}_{out}\right) - x\right.$$

$$\left.\sin\left(\text{angle}_{out}\right)\right)$$

In OpenSCAD, the only parameters that change from the red light case are

```
lambda = 2 * 480 / 480;
refractive_index = 1.509;
```

Figures 8-13 shows these models side-by-side, with the longer wavelength (red) light on the left, and the violet on the right. You can clearly see the light turning as it goes into the glass (lower half of the model), and the wavelengths getting shorter as the wavefront turns. However, the difference in how they turn (how much the blue will turn away from the red) is less than a degree, so it is pretty much invisible in these models. We can compute it and see that the difference in angle in the medium is about two tenths of a degree, 27.7° versus 27.9°.

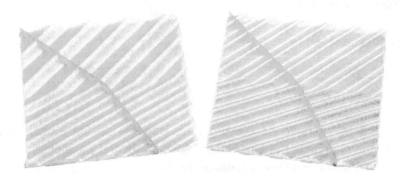

FIGURE 8-13

Red light (left) and violet light (right) going from air to glass

Let's return to our example of cars passing us on a road, now a many-laned highway. Think of a line of cars driving next to one another across all the lanes and matching their speeds. If the speed limit suddenly dropped, the cars would have to slow down. If there was another row of cars behind them, the second row would slow down later, and thus end up closer to the first line as they both drive at the lowered speed.

Suppose that, say, that first the rightmost lane needed to drop to a speed 20 miles an hour less than the next lane. Then a few seconds later, the lane next to it also needed to drop and so on, all the way across this line of cars that had been driving alongside one another.

Since the cars in the left lanes would continue at their higher speed longer, they would get ahead of their neighbors. When all the cars were in the speed-restricted zone, the line of cars would now cross the road diagonally, with the left-most cars ahead of those farther to the right, even though they are all still traveling in the same direction as when they started. This is what is going on with our light waves in Figure 8-13.

To exaggerate the effect so we can see it, we will invent a glass made of pure unobtainium (a mythical substance) which has an index of refraction of 1.6 for wavelengths of 2 in our arbitrary model units, and an index of 1.2 when the wavelength is 4. (Glass has a slight difference in index of refraction between red and violet light too, but it is a tiny difference as we saw in the last example). Having the wavelengths as a multiple of each other will make it easier to do comparisons on the incoming waves. In other words, we ran `refraction.scad` with the same equations as we did for red and violet, but with

```
lambda = 2;
refractive_index = 1.6;
```

for the first case, and

```
lambda = 4;
refractive_index = 1.2;
```

In Figure 8-14, we have these two models side by side. Once again you can see the change in the angle of the wavefronts and wavelength after going into the new medium. However, as shown in Figure 8-15, we can now line these models up and backlight them to see the difference between our two wavelengths.

FIGURE 8-14

Longer wavelength (left) and shorter (right) refracting

FIGURE 8-15

Lining up the two models in Figure 8-14.

Our model shows the waves, as well as the normal to the waves. Most books show the normal to the waves as a ray and calculate with it instead of thinking about the waves. You can convince yourself that the angle of the waves to the interface of the medium is the same as the angle of the normal to the wave with the normal to the medium.

Finally, Figure 8-16 shows how light is split up by a prism into a rainbow. The light coming out of the prism is turned twice, on entering and on exiting the prism, which exaggerates the turns further. Note the angles at which the light enters and (after passing through two interfaces) the rainbow emerges. Putting the sunlight or flashlight beam entering through a slit or other narrow opening will make the effect more pronounced and easier to see. (We did this in creating Figure 8-16). This figure was created with an LED flashlight which does not produce the full spectrum of colors; try to use an incandescent source if you can.

FIGURE 8-16

Prism splitting light into its component colors

Lenses

Another application of refraction is manipulating light with *lenses*. Lenses (like those in eyeglasses, or a magnifying glass) can bend and collect light in ways that let us see things that are far away or very small. Lenses are made of glass or some other transparent material. When light enters and exits the curved surface of a lens it is bent relative to the light entering it.

Lenses fundamentally fall into two categories: converging (with convex sides) and diverging (with concave sides). Converging ones, like a magnifying glass, will take in light and focus it at a point on the other side. Diverging ones will make entering light waves spread apart.

Lenses and their applications could fill a library's worth of books, and there are many experiments to do with them. We would stray far from our trigonometry mission if we went too far down that path, though, so we will stop here and suggest other routes you might want to pursue on your own. First you might read the Wikipedia article "Lenses" for an overview. Then search online for "lens experiments" to either watch videos of people with professional equipment trying out these principles, or perhaps simpler do-it-yourself projects.

Putting together several lenses strategically is the basis of a telescope or a microscope. If you would like to create an inexpensive microscope, you might try creating a **Foldscope** (https://foldscope.com). These are microscopes made from paper plus some inexpensive lenses and can be purchased with a variety of accessories.

Buying a kit with lenses, as we also noted in our Chapter 7 discussion of sextants, is probably easier than trying to source all the pieces yourself. There are also assembled inexpensive microscopes and telescopes available, if you are more interested in the application than in experimenting with lenses. Another option is to buy an add-on lens kit that enhances a phone camera.

Reflection

Next, let's think about what happens when light waves hit a mirror and are reflected. If we are directly in front of a mirror, we see ourselves (but reversed in the mirror). If, on the other hand, we shine a light beam at a mirror at an angle, it will reflect off at the same angle relative to a line perpendicular to the incoming beam (Figure 8-17). We used that fact in Chapter 7 when we investigated the design of a sextant.

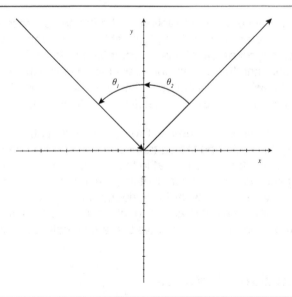

FIGURE 8-17

Light reflecting off a mirror

Images in a mirror appear as they do because rays on the right reflect back on the right and those on the left on the left, so it appears reversed to the viewer. It is, however, how someone standing with their back to the mirror would see you. Try this simple experiment. Stand next to a mirror with a friend. Look into your friend's eyes (in the mirror). To do that, your eyeline will need to be pointed at the centerline of the two of you, following the rays in Figure 8-17.

Reflection is a surprisingly complex phenomenon. If we want to think about the phase of the waves (as opposed to their angle incident on a surface) we need to look at Stokes' Relations (see sidebar). There are many other wave phenomena we could discuss and model, too. Our 2016 book, *3D Printed Science Projects*, has a chapter "Light and Other Waves" which goes in somewhat different directions than this book does. In that book, we look at light interfering after it has gone through two slits, which has some interesting properties.

Stokes' Relations

Sir George Gabriel Stokes, an Irish-born mathematician and physicist, was the Lukasian Professor at Cambridge University in England from 1849 to 1903. This was the position held earlier by Isaac Newton, and later by Stephen Hawking. Stokes came up with rules about how the phase of a wave changes when it is reflected. For a case where light intersects a material with higher index of refraction, the phase will change 180° for the reflected wave.

Its angles of incidence and of refraction will be the same phase as each other, but the reflected wave will be phase shifted 180° from the incident one. These are called "Stokes' Relations" or sometimes "Stokes' Law." (He was a busy guy, though, and there is another calculus relationship more commonly called "Stokes' Law".) We will not delve into the physics here, but if you are interested you can look at "Stokes' Relations" in Wikipedia.

A plane wave reflecting off a surface, like light bouncing off a wall, can be quite complex, depending on the surface. Some light may reflect, some may continue on but be altered somewhat in the new medium, as we discuss in the upcoming section on refraction. Not all reflected waves change phase. Delving into this would take us far from trigonometry so we will stop here but suffice it to say that the electric and magnetic components of the wave may interact with the surface in various ways that affect their respective phases.

Water and Sound Waves

In the last section, we learned that electromagnetic waves are created by electric and magnetic fields propagating at right angles to each other. This means that they can spread quite happily in a vacuum (in fact, are fastest there), as we saw in our refraction discussion. The sine waves representing the two fields oscillate in a plane perpendicular to the direction of transmission of the wave, as we saw, for instance, in Figure 8-15 and other similar models in this chapter. Waves like this are called *transverse waves*.

Surface water waves are also transverse waves. A ripple in a pond will spread out with oscillations in the vertical direction. Our models for electromagnetic waves with a source, a plane wave, and overlapping in Figures 8-4 through 8-6 apply to the similar situation to waves on the surface of water as well.

Imagine a buoy floating on the surface of the water far from shore. It will not be carried forward as the water wave oscillates. It will bob up and down and make a small circle at right angles to the waves as the wave passes through its volume of water. Water waves are also *dissipative*, which means that the energy that created them will over some distance or time peter out.

A light wave in vacuum can in principle go forever and is nondissipative. However, as a wave spreads out, you can imagine its farthest surface forms a sphere that keeps expanding. Since the energy put into the wave at its source is a constant, the energy will drop off with the surface area of that sphere, which is proportional to $\frac{1}{r^2}$, where r is the radius of that

sphere (or, if you prefer, the distance of an observer looking at the source of the light).

Sound waves are a bit different. They also need a medium to transmit them (hence no one being able to hear you scream in the vacuum of space). They are called *longitudinal waves*, which means the oscillations that create them move back and forth as periodic increases and decreases in pressure of the air or other medium they are traveling through. You can think about sound as being ripples in the air (like our other models), but it is not really accurate.

Modeling a Helicoid

In this chapter so far, we have thought about using waves to model physical phenomena. Sines and cosines are often used to model other purely mathematical constructions as well, particularly periodic ones. Back in Chapter 4, we generated a surface called a helicoid. It is generated by filling in a spiral-staircase-like shape called a *helix*. We revisit it now because we can write very simple equations for a helix's Cartesian x, y and z coordinates in terms of sinusoids:

$$x = \cos(z)$$
$$y = \sin(z)$$
$$z = z$$

Defining variables in terms of functions of one or more others is called creating a *parametric set of equations*. The way we implemented this in OpenSCAD was to imagine that we were drawing out a circle, like creating the unit circle in Chapter 5. Once again, we start with a fixed-length line along the x axis with one end at the origin, and we sweep it counterclockwise. The end at the origin is the pivot.

This time, though, as we are sweeping around, we are also stepping upwards instead of creating a 2D circle. The model `sinusoidhelicoid.scad` accomplishes this with OpenSCAD's `linear_extrude` function, which takes a line and rotates it through a number of degrees specified by the parameter `twist` (360° in this case). The model also prints out x and y axes to help orient the model (Figure 8-18) and to act as a stand for it.

FIGURE 8-18

Helicoid

This OpenSCAD model itself is relatively simple. First it sets a few parameters, and then creates the axes, followed by rotating and extruding the helicoid and a central cylinder that supports it. Here it is in its entirety:

```
size = 18; //length of the line
base = 100; // length of the x-y axes
// variables that set surface quality
$fs = .2;
$fa = 2;
// create the base
linear_extrude(5) {
  square([5, base], center = true);
  square([base, 5], center = true);
}
//sweep the line through twist = 360 degrees
translate([0, 0, 5]) linear_extrude(size * 2 * PI, twist = 360) hull() {
  circle();
  translate([size, 0, 0]) circle();
}
// make a central post to support the helix.
cylinder(r = 2.5, h = 5 + size * 2 * PI);
```

We can also look at the model in the x-z plane and see a cosine wave, and in the y-z plane, a sine wave (Figures 8-19 and 8-20) consistent with the parametric equations.

FIGURE 8-19

Helicoid seen in the x-z plane

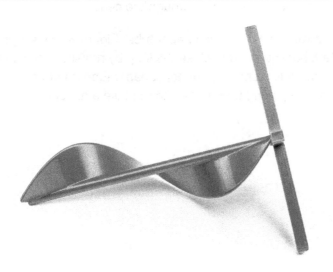

FIGURE 8-20

*Helicoid seen in the
y–z plane*

A helicoid is a *minimal surface*. If we had a version of this model consisting of just the outside edge and central support, and dipped it into a bubble solution, we would get a soap film the same shape. Take some wire (16-gauge uninsulated craft wire works well) and make a helix with a piece of wire connecting its ends through the center.

Wrapping the wire around a toilet paper tube (Figure 8-21) is a good way to get started, but you need to close the loop by running a piece of wire down the center after removing the toilet paper tube (Figures 8-22 and 8-23). Leave a bit poking up on one end to make a handle.

FIGURE 8-21

Wire helix starting around a toilet paper tube

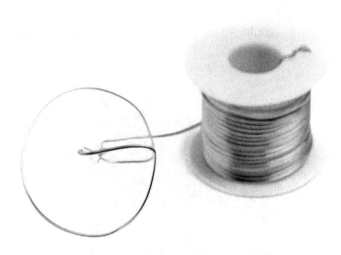

FIGURE 8-22

After sliding the wire off the tube

FIGURE 8-23

Closing the loop

To create a helicoid (Figure 8-24) fill a bucket or other big-enough container with some soapy water and dip in the wire helix. Gently pull it out of the water, and with luck you should have a helicoid surface filling in your helix.

FIGURE 8-24

Soap bubble helicoid

If you were wondering, helicoidal structures occur in many natural systems (most of them microscopic). For example, some beetles have them in their wings, which then reflect circularly polarized light differently depending on orientation. They can also be found inside certain kinds of living cells, in the shape of the organelle called the endoplasmic reticulum.

Chapter Key Points

In this chapter, we explored properties of sinusoid continuous curves, and the implications of those properties for real physical systems. We explored electromagnetic waves, like light, and the properties of reflection and refraction. We also looked at an example using sine and cosine waves to model more-complex curves, like a helicoid.

Terminology and Symbols

Here are some terms from the chapter you can look up for more in-depth information:

- constructive and destructive interference
- electromagnetic waves
- helicoid
- incident angle (denoted by Greek letter theta θ, with subscripts for before and after an interface)
- index of refraction (denoted by n, with subscripts for before and after an interface)
- lenses
- longitudinal wave
- Nyquist frequency
- prisms
- sawtooth wave
- Snell's Law
- square wave
- Stokes' Relations
- transverse wave
- triangle wave
- water waves
- wavelength (denoted by the Greek letter lambda λ)

References

Here are some books and other references if you want to dig further into the material in this chapter:

Forbes, N., and Mahon, B. (2021). *Faraday, Maxwell, and the Electromagnetic Field.* Tantor Audio (Audiobook), also 2019 paperback from Prometheus books. Very readable story of Faraday and Maxwell's experiments, including their dead ends.

Horvath, J., and Cameron, R. (2106). *3D Printed Science Projects.* Apress. The chapter "Light and Other Waves" also looks at some basic wave phenomena.

Some of the phenomena in this chapter are easier to view as an animation. Some good mathematical visualization channels you can search for on YouTube are those of Nils Berglund and 3Blue1Brown, although the latter channel might be a little esoteric sometimes for the readers of this book.

Ellipses and Circles

3D Printable models used in this chapter

See Chapter 2 for directions on where and how to download these models.

- `conic_sections_set.scad`
    - This model slices a cone to produce all the conic sections.

- `ellipse_slider.scad`
    - This model creates an ellipse given major and minor axes, with holes at the focus and a slider around the outside.
- You will also need:
    - Two pushpins
    - A piece of foam core or thick cardboard
    - A piece of thin and not stretchy cord, like sewing thread, about a foot
    - Elastic cord. We used 0.8 mm elastic cord, also called beading thread or crafting cord. You will need about 1 foot per model.
    - If you want to make a Play-Doh alternative cone:
        - 8 ounces Play-Doh or other modeling clay
        - Unwaxed dental floss (to cut the Play-Doh)
        - Parchment or wax paper
        - A shallow box, and something to use as a spacer (see description of Play-Doh cone creation for requirements)
        - A friend to help hold things still

This book started out with simple trigonometry of individual triangles, and then used an analysis of the unit circle to see how to generate continuous sinusoidal curves. Now in these next three chapters, we move on to *analytic geometry*, which in some similar ways ties together simple physical models and the equations and graphs for idealized versions of them. In particular, we look at the results of slicing a cone, the *conic sections*.

The Greek mathematician and astronomer Apollonius of Perga lived about 2100 years ago. He and his contemporaries figured out that slicing (or "sectioning") a cone in different ways created shapes bounded by interesting curves: *circles*, *ellipses*, *parabolas*, and *hyperbolas*.

Apollonius understood many of the features of these curves but did not have a good way of writing them down in equations. That had to wait until Descartes came up with his coordinate systems in the early 1600s. It took another hundred years or so for Euler to come along and write the equations for ellipses, parabolas, and hyperbolas as we do today. For more on this history, see Uta Merzbach and Carl Boyer's book we have noted in earlier chapters, *A History of Mathematics*. They go into quite a bit of detail about what Apollonius knew and how he knew it.

In these next few chapters, we are going to start off like Apollonius to get an intuitive understanding of conic sections, and try to end up like Euler, with equations defining these curves. Our book *Make: Geometry* has a much longer development of the basics of conic sections. There is a condensed excerpt in this chapter.

The *Make: Geometry* model repository also has a model that allows a single arbitrary cut through a solid. If you need a bit more background or want to play with some additional 3D printable models, you might want to start there. Our emphasis here will be on building intuition leading to graphing and interpreting conics and the equations that describe them.

In this chapter, we give a general introduction to all the conic sections. We then dive into the ellipse and circle in more depth. In Chapter 10 we look at parabolas and talk about the math needed to analyze the *quadratic equations* that describe them. Finally, in Chapter 11, we explore hyperbolas. Along the way we will point out where (and why) you are likely to meet these curves in real life.

Conic Sections

The *conic sections* resulting from slicing through a cone with a plane are circles, ellipses, parabolas, and hyperbolas. (Hyperbolas require two cones arranged point-to-point, but we will get to that later.) As we will see, we can predict, based on the angle of the cut, which of the conic sections we will get.

Anatomy of a Cone

The *slant angle* of a cone (Figure 9-1) is the angle the side of the cone makes with its base. If we have the height, h, of the cone, and the radius of the base, r, we know that the tangent of the slant angle has to be the height divided by the radius.

$$\text{Slant\_angle} = \tan^{-1}\left(\frac{\text{height}}{\text{radius}}\right)$$

FIGURE 9-1

Slant angle and slant height

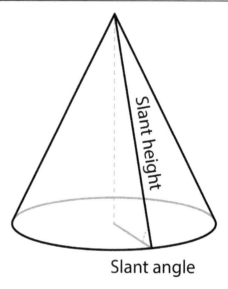

Slant angle

Note that some books measure the angle of the cutting plane relative to the *vertical* axis of the cone, so be sure and know what convention a book is using if you read about conic sections elsewhere. We measure the angle from the base of the cone.

The *slant height*, also illustrated in Figure 9-1, is the distance from the base of the cone to the vertex, along a side. It is the hypotenuse of the triangle made by the base and the height. From the Pythagorean Theorem,

$$\text{Slant\_height} = \sqrt{r^2 + h^2}$$

Conics Overview Model

The OpenSCAD model `conic_sections_set.scad` prints a cone with cuts that create all the conic sections. You might need a bit of double-sticky tape or museum putty to hold this model together, depending on the slipperiness of the filament you are using. For the photos in this chapter, we printed two sets using different color filaments and re-assembled the cones while alternating the pieces to make the cuts more visible (Figure

9-2). The two sides of the cut are beveled on one side and have a lip on the other. This makes it clearer where the cut is and helps them hold together better when you assemble the set. An idealized mathematical cone would not have these bevels.

FIGURE 9-2

The set of all the sections (model printed twice)

The model has the radius of the cone and its height equal to each other, and we do not recommend changing any parameters since the relative arrangements of all the cuts are carefully calibrated so as not to cross each other. This model is best scaled in your slicer if it needs to be bigger or smaller. Be sure that you scale all axes by the same amount.

If you do not have a 3D printer, create a cone out of Play-Doh and cut it at appropriate angles. Dental floss works well to make clean cuts. Plan ahead and mark ALL the cuts before you try it to make sure they will not intersect. We have also seen people buy a Styrofoam cone (often used by florists and chefs for fancy centerpieces) at a craft store and cut it by some means. Our experiments, though, indicate this will be very messy and lead to unsatisfactory results, at least if classroom-safe cutting tools are used. We summarize the process of cutting sections of a Play-Doh cone in a later section of this chapter.

Circle Cross-section

Now, let's start looking at the parts of our 3D printed model. If we cut through a cone with a plane parallel to its base, the cut we make is a circle. How big a circle it is depends on how far up from the bottom of the cone we make the cut. But any cut parallel to the base will be a circle (Figure 9-3). Right at the vertex of the cone, the cut would be a point, but that is a special case.

FIGURE 9-3

Both sides of the circular cross-section cut.

Ellipse Cross-section

If we now take our cone and make a cut at an angle to the horizontal that is more than zero, but less than the slant angle, we get an *ellipse* — a circle stretched out in one dimension (Figure 9-4). Ellipses come up all the time in astronomy; the orbits of the planets around the Sun and of the Moon around the Earth are ellipses. We see how to derive an equation for an ellipse later in this chapter.

FIGURE 9-4

Both sides of the ellipse cut

The more we tilt the cutting plane relative to the base of the cone, the longer and skinnier the ellipse gets. Like the circular cross-sections, the ellipse will be larger if the cutting plane does not pass very close to the vertex. It is obvious why we get an ellipse if the cutting plane is at an angle to the horizontal greater than zero since cutting at a slant elongates one axis. It is less obvious why the slant angle is the upper bound. Let's talk about the parabola to see where that constraint comes from.

Parabola Cross-section

What happens if we cut through the cone parallel to its side (that is, when the cutting plane is at the slant angle)? The resulting cross-section is a curve sort of between the letters U and V in shape, called a *parabola* (Figure 9-5). Parabolas make their appearance in many physical situations, notably in parabolic dish antennas or mirrors that concentrate light in one spot called a *focus*. In Chapter 11, we will talk about why those systems work. For now, let's just see what they look like and how they arise from cutting a cone.

As we saw earlier in the chapter, the slant angle is

$$\text{Slant\_angle} = \tan^{-1}\left(\frac{\text{height}}{\text{radius}}\right)$$

In this case, since the radius of the cone and its height are the same,

$$\text{Slant\_angle} = \tan^{-1}(1) = 45°$$

FIGURE 9-5

Parabola slice when cone height and radius are equal

Unlike an ellipse or a circle, the parabola is not a closed curve. Since our physical cone is finite it cuts off at the bottom of the cone, but an ideal cone goes on to infinity. The straight part of the cross-section along the bottom of the cone is not considered part of the parabola. Planes cutting at the slant angle (and more steeply) create curves that are not closed (parabola and hyperbola.) You can see why if you look at the parabola's cross-section: any cut steeper than that would always be cut off at the bottom of the cone. If the cone was infinite, the sides of the cut would go on forever.

Hyperbola Cross-section

Finally, if we make a cut steeper than the slant angle, we get a *hyperbola*. Technically, what we can see in Figure 9-6 is only half (one branch) of a hyperbola. To get both branches, we need to start with two cones vertex to vertex and allow the cutting plane to go through both cones (but not through the vertex point). We explore hyperbolas in depth in Chapter 11.

FIGURE 9-6

One branch of a hyperbola

Cutting a Play-Doh Cone

If you do not have access to a 3D printer, a Play-Doh cone is a pretty good substitute. To cut a parabola from a dough cone, first create a cone. We used 8 ounces of Play-Doh. The parabola and circle cuts have to be at one particular angle and the ellipse and hyperbola have many options, so make the parabola cut first.

Find a shallow box and put in a spacer (we used a stack of sticky notes) with a sheet of parchment paper on top. Then put the cone on top of that so that the cut you want to make can use the lip of the box as a guide (Figure 9-7). Since the parabola cut is supposed to be parallel to a side, you will want to make your cut parallel to the table.

Get a friend to gently hold the cone steady since it will want to slosh around (Figure 9-8). Use unwaxed dental floss or another appropriate implement to make the cut. As with the 3D printed version, you will get a curve that is open on the bottom (Figure 9-9).

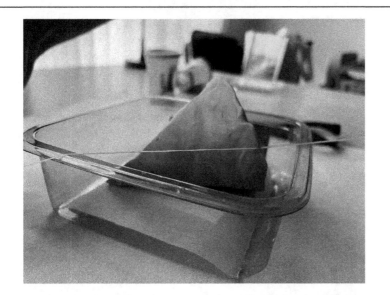

FIGURE 9-7

Setup before cutting

FIGURE 9-8

Making the parabola cut

FIGURE 9-9

The parabola cut

If you want to, you can add cuts to make an ellipse and one branch of a hyperbola as well. Those can be at any convenient angle as described in their respective sections (and summarized in the table in the next section). The quality will not be great since each cut will tend to mush the preceding ones a bit, but you can get the idea and experiment.

Conic Sections and Angles

In summary, then, we can predict what cross-section we will get by making a slice across the cone from the angle of the cutting plane, as shown in this table 9-1.

Table 9-1. Conic section angles

Section	The angle of the cutting plane to the base of the cone (not through the vertex of the cone)
Circle	0° (parallel to the base of the cone)
Ellipse	More than 0°, less than slant angle
Parabola	Equal to slant angle
Hyperbola	Greater than the slant angle, up through a maximum of 90°

Ellipses and Circles

Now that we have the big picture of conic sections, we can dig into their properties and get a little intuition about how they work. In this chapter, we focus on the circle and ellipse. We can define a circle as a curve that is a constant distance from one point. Imagine pinning a piece of string down at one end so that it moves freely. If you swing it around that pinned point, the end will trace out a circle. Drawing an ellipse requires two pins instead of one.

Take a piece of cardboard or foam core, thick enough so that pushpins will not poke all the way through. (We used two layers of a shipping cardboard box). Tape a piece of paper over this base. Take two pushpins and use them to pin down the ends of a piece of string (Figure 9-10). You can tie a knot in the string to prevent it from unraveling and make a more solid anchor for the pushpins. The string should be loose. About 25% longer than the distance between the pins should be about right. String used to tie up packages works, or you can use thinner string if you prefer.

FIGURE 9-10

Setting up the string to draw an ellipse

Now, put your pencil alongside the string, and stretch it taut, but not so tight that the pushpins want to pop out (Figure 9-11).

FIGURE 9-11

Placing your pencil

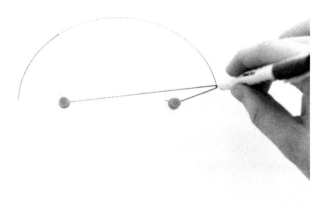

Now, keeping the string taut (and probably using your other hand to hold the pushpins down) you can move the pencil around the pushpins. You should be able to draw the top or bottom half of an ellipse(depending on where you started), and then adjust your pencil to make the other half (Figure 9-12). Finally, you will have the entire ellipse (Figure 9-13).

FIGURE 9-12

Drawing the bottom of the ellipse

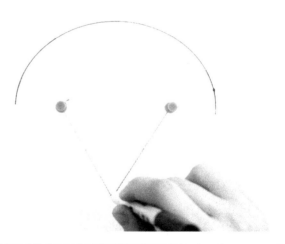

FIGURE 9-13

Finished ellipse

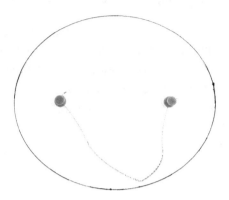

Try moving the pushpins farther apart. The ellipse will get longer and skinnier (Figure 9-14). On the other hand, if you imagined putting them right on top of one another (or, more likely, using only one push pin), you would just have a circle.

FIGURE 9-14

Moving the pushpins farther apart

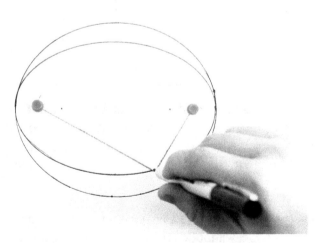

A circle is defined as a curve that is a constant distance from its center. That distance is called the *radius* of the circle. You can think of an ellipse as a more general type of circle that is stretched out along one axis. An ellipse is a curve that is a constant sum of distances from two points, called *foci,* pronounced "foe-sigh" or "foe-kigh". (Each one individually is called a *focus*). In the drawing we just did, the pushpins are the foci of the ellipse.

An ellipse has two numbers that characterize it. The longest dimension of the ellipse is called the *major axis.* Half of that is called the semimajor axis, usually referred to by the letter a. The shortest dimension is called the minor axis, and half of it (the semiminor axis) is usually called b. The foci lie on the major axis. They are both the same distance from the center of the ellipse. That distance is called c and is dependent on a and b in a way we will describe later (Figure 9-15).

FIGURE 9-15

Semimajor and semiminor axes and foci of an ellipse

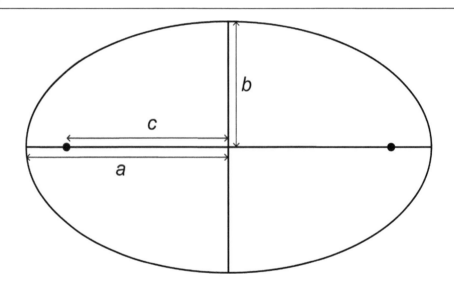

You can also use a loop of string, instead of pinning the string (Figure 9-16). This has the advantage that it lets you run your pencil all the way around the ellipse without picking it up, but the math is just a bit more complicated. In this case, the circumference of your loop would be $2(a + c)$.

If you want to see how the ellipse changes when you adjust the distance between the foci without changing the major axis, you will need to adjust the loop so that its circumference remains $2(a + c)$.

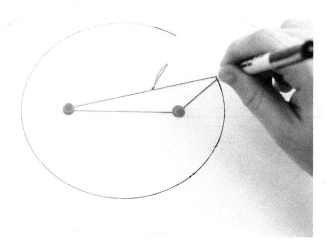

FIGURE 9-16

The loop of string variation

For any point on an ellipse, the sum of the distances to the two foci remains constant. The dashed lines in Figure 9-17 show the distances from the two foci to a point at the end of the semiminor axis. No matter which point we choose on the ellipse, the sum of these two lengths remains the same. This is why we can draw an ellipse with two fixed points and a fixed length of string. You can follow the rest of this discussion with your string and pushpin model, but we also created a 3D printed model, `ellipse_slider.scad`, which works on the same principle and is a little easier to play with than our pushpins and string.

The model prints in two parts: an ellipse with two biggish holes for the foci, and a smaller part that slides around the outside of the ellipse. You will need some elastic cord; we found that 0.8 mm elastic cord (shown in the photos in this chapter) works well. As it happens, this is more or less the stretchy cord used to hold low-cost conference badge lanyards. If you have a chamber-of-commerce event badge lying about you can scavenge that.

The length of the string should not change once it is in place, so it would technically be more correct to do this with a string that does not stretch. However, we found that it was difficult to tie tightly enough to keep everything in place. The length of the string is twice what we would need to draw the model with two pushpins, since it wraps around both sides of the model.

FIGURE 9-17

Ellipse dimensions

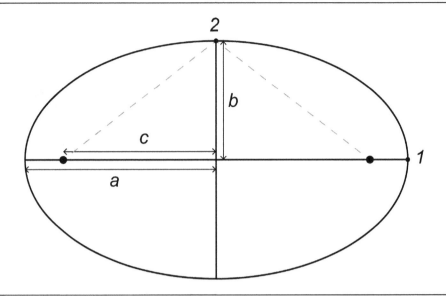

The model has a few parameters you may want to change:

- d = [80, 60];
    - Major and minor axes, in mm (in other words, [2*a, 2*b])
- holesize = 5;
    - Diameter of the focus holes, mm
- h = 8;
    - Thickness of the piece, mm

Loop the cord through in a figure-8 pattern, crossing over inside the slider and through each hole. The string will be going from the two foci to the slider on both sides of the models, so you will need enough string for that. Building it is a little tricky. First, print the model. For this set of photos, we printed it hollow with a translucent material, to make it easier to see how we are routing the cord behind it. Next, cut a piece of elastic cord a bit more than twice as long as the major axis. Put the cord through one of the holes at the focus, from front to back, then through the slider. This piece of cord will be on what we will call the back of the ellipse (Figure 9-18).

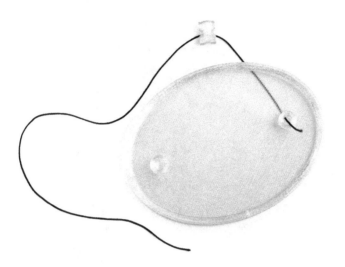

FIGURE 9-18

*Starting to string the
cord on the model*

Then, run the cord through the other focus, again going from the front to back (Figure 9-19). Then go through the slider again, from the back to the front (Figure 9-20).Tie off the cord on the front of the model, then you can slide the string through to hide the knot inside one focus. You should now have cord going from the foci to the slider on both sides (Figure 9-21).

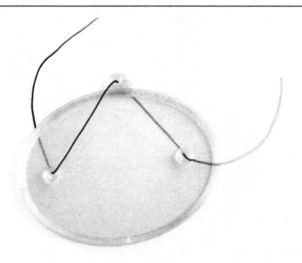

FIGURE 9-19

*Starting to string the
cord on the model*

FIGURE 9-20

*Bringing the string
around the back*

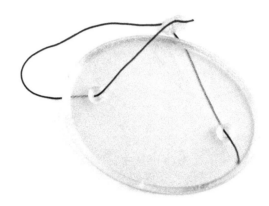

FIGURE 9-21

*Tie off the string to
finish the model*

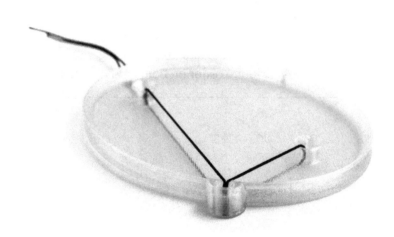

Eccentricity

All ellipses, regardless of how far apart their foci are, have the property
that the sum of the distances from any point on the ellipse to its two foci
is a constant. As a practical matter, you cannot really have the foci much
closer than 5 mm apart with this model because then the holes would
merge. If the ellipse is too skinny, the holes for the foci will fall off the
ends since they are quite large. Experiment a little in OpenSCAD to find

some good test cases. Figure 9-22 shows a nearly circular ellipse, using the parameter `d = [80, 79]`. Figure 9-23 is a long skinny ellipse, with `d = [80, 45]`.

FIGURE 9-22

Nearly circular ellipse

FIGURE 9-23

Thinner ellipse

If you were to put the two foci together, they would become one hole in the model — and you would get a circle with a constant radius. For example, using `d = [80, 80];` will work. How "circular" or how "long and skinny" an ellipse is has a metric called the *eccentricity*. (Not to be confused with the description of that relative who only shows up on major holidays.) The eccentricity of an ellipse is usually symbolized with a lower-case e, and is defined as

$$e = \sqrt{1 - \frac{b^2}{a^2}}$$

Note that if a and b are equal (that is, if we have a circle) the eccentricity, e, of a circle is zero. The eccentricity of a parabola and a hyperbola are defined by different formulas, which we talk about in later chapters.

Finding the Foci

It would be good to know where the foci are, in case we want to design something elliptical and draw it. The easiest way to think about finding the foci is to imagine an ellipse and to think about two special cases. If we can prove a relationship for these special cases, we can then reason about how to generalize that case to be sure it will always work.

The distance from the center of the ellipse to each focus is c. The length of the semimajor axis is a and the semiminor, b. Let's look at two special cases with our ellipse model. Pull the slider on the model all the way to the left, so that the string from both foci lies along the longer (major) axis of the ellipse (Figure 9-24).

FIGURE 9-24

Finding the long axis of the ellipse

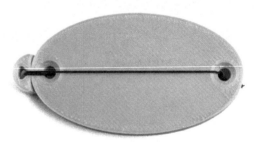

If the distance from the focus on the left is

$$a + c$$

then the distance from the focus on the right is

$$a - c$$

The sum of the two is

$$a + c + a - c$$

or just $2a$. The sum of the distances from the foci to any point on the ellipse is a constant. Now that we know that this sum of distances is $2a$ at one point, we can infer that this is true for every point on it.

But that still does not tell us how the semiminor axis, b, comes in. By analogy, let's try thinking through the point at one end of the semiminor axis. (At the top, if you have the major axis running left to right, as in Figure 9-25.)

FIGURE 9-25

Finding the shorter axis of the ellipse

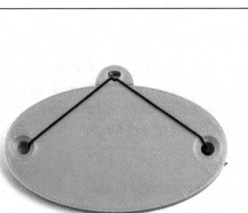

We know that the distance from a focus to the center of the ellipse is c, by definition. Likewise, we know that the distance from the center of the ellipse along the semiminor axis is b, by that definition. So, we can make a right triangle that has one side of length b, one side of length c, and the hypotenuse is the distance from the focus to the point at the end of the semiminor axis. Pythagoras tells us that the length of this hypotenuse is (Figure 9-26) $\sqrt{c^2 + b^2}$.

FIGURE 9-26

Ellipse dimensions

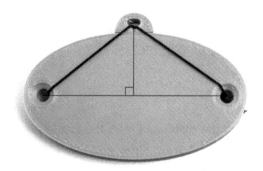

Now, that hypotenuse is only the distance from one focus to our point in question. The situation is symmetrical, so the sum of the distances at that point is $2\sqrt{c^2 + b^2}$.

We also know from what we just figured out with the point at the end of the semimajor axis that this distance also must be $2a$. We can set them equal to each other, since we know that this sum of distances is the same no matter what point on the ellipse we pick.

$$2\sqrt{c^2 + b^2} = 2a$$

Divide by 2 and square both sides to get

$$c^2 + b^2 = a^2$$

or

$$c^2 = a^2 - b^2$$

If you are used to looking at the Pythagorean Theorem that may bother you, but in math the letters a, b, and c are used for many things. Unfortunately, in this case, it is in a way that fights our Pythagorean instincts from earlier chapters.

As an exercise in getting some intuition about these models, take a ruler and measure the semimajor and semiminor axes on the printed-out ellipses, and calculate where the foci should be. Measure to the centers of the holes in the models, or measure to the edge of one hole and add the model variable `holesize` to the result (mm) to allow for the hole. This should be $a - c$. See if the values agree and consider some extreme cases.

Equation of a Circle and an Ellipse

Finally, is there an equation we can write that describes an ellipse in terms of x and y coordinates? We can also use the Pythagorean Theorem to find the equation that describes an ellipse. First, let's see if we can find the equation for any point on a circle so that we will have some intuition about what sort of equation to expect, since they should be similar.

Say we have a circle that has a radius, r, and that its center is where the x and y axes cross. Pick any random point on the circle (x, y) and draw a triangle with one side on the x axis, and one parallel to the y axis, with the hypotenuse being the radius. If we let the radius equal 1, we get right back to our unit circle in Chapter 5. There, we used this diagram to figure out sine and cosine as we went around the circle. But for now, let's continue with this more general case in which the radius does not have to be equal to 1 (Figure 9-27). Pythagoras then tells us that

$$x^2 + y^2 = r^2$$

which is the equation of a circle that has a center at $x = 0$ and $y = 0$, with radius r.

FIGURE 9-27

A circle on the coordinate plane

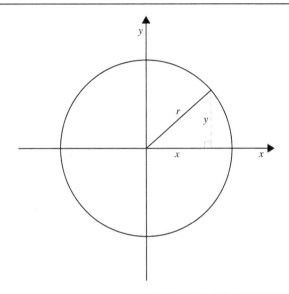

Since that was not so bad, let's press our luck and try the same thing with an ellipse, also centered at $x = 0$ and $y = 0$. But the ellipse goes from $-a$ to $+a$ along the x axis, and $-b$ to $+b$ along the y axis.

If we divided all the x values by a, and all the y values by b, our ellipse would scale back into being a circle, with a radius equal to 1. So now, we can do the very same thing: draw a triangle with one side being of length $\frac{x}{a}$, and another of $\frac{y}{b}$ (Figure 9-28) to see that the equivalent equation of an ellipse is

$$\frac{x^2}{a^2} + \frac{y^2}{b^2} = 1$$

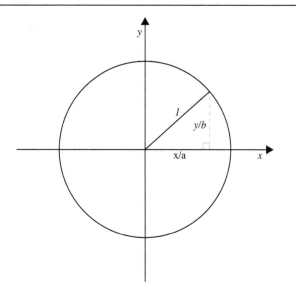

FIGURE 9-28

A circle (a scaled ellipse) on the coordinate plane

This is the equation for an ellipse with its major (longer) axis along the x axis, its minor (shorter) axis along the y axis, and its center at $x = 0$ and $y = 0$. By convention, the longer axis is always of length a, regardless of orientation, and b is the length of the shorter one.

Translating Circles and Ellipses

Translating a curve just means that we shift it in the x or y coordinate, or both. Sometimes it is handy to get the equation for an ellipse (or circle) that is not right on the axis. If we want to offset an ellipse in x and y coordinates, for example, all we do is subtract off the offset in the relevant direction. Then we get right back to the same curve we started with. So, for instance if we wanted an ellipse centered at $x = 5$ and $y = -3$, the equation for it would be

$$\frac{(x-5)^2}{a^2} + \frac{(x+3)^2}{b^2} = 1$$

Note that we subtract off the amount we have shifted, so shifting -3 means adding +3 in the y dimension. The meanings of a and b remain the same (lengths of the axes). Figure 9-29 shows this shifted ellipse.

FIGURE 9-29

An ellipse centered at (5, -3)

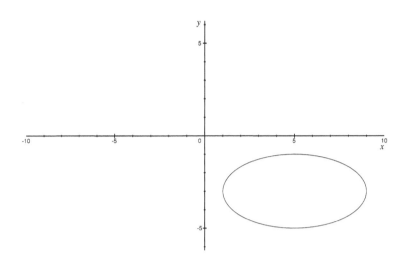

Equations of Rotated Circles and Ellipses

Obviously, rotating a circle about its center has no effect, since circles are the same curve no matter how you rotate them. Ellipses, however, can be rotated the same way we rotated our waves in Chapter 8: by working in a rotated coordinate system. We saw that if we rotated our coordinates by an angle A, we could transform from the rotated (X, Y) coordinates to the original (x, y) by using the relationships

$$X = x \cos (A) + y \sin (A)$$

$$Y = y \cos (A) - x \sin (A)$$

So, if we have an ellipse that has been rotated an angle A, in the rotated coordinates it is just

$$\frac{X^2}{a^2} + \frac{Y^2}{b^2} = 1$$

and in our original coordinates it would be

$$\frac{[x \cos (A) + y \sin (A)]^2}{a^2} + \frac{[y \cos (A) - x \sin (A)]^2}{b^2} = 1$$

where we put the equivalent coordinates in square brackets to make it easier to see what we are doing. You can multiply it out and collect terms if you like but suffice it to say that you should try to work with ellipses aligned with your coordinate axes if you can! Figure 9-30 shows this situation; compare it to Figure 8-10 which showed us how to change coordinates to rotate our incident and refracted waves.

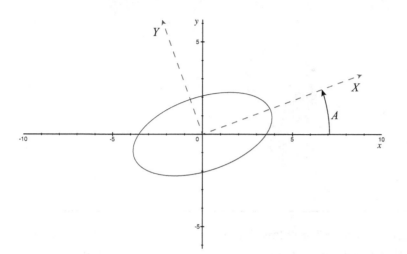

FIGURE 9-30

A rotated ellipse (centered at (0, 0))

Area of an Ellipse

We know the area of a circle of radius r is πr^2. If we draw a square around this circle so it is just touching the walls of a square, the area of the square will be four times the radius squared (since the radius is only half length of the side of a square). Thus, the area of the circle is $\frac{\pi}{4}$ times the area of the square, or about 79% the area of the square (Figure 9-31). This circle is called the *inscribed* circle, or the biggest circle that can fit inside the square.

FIGURE 9-31

An inscribed circle

What about an ellipse? Imagine we stretch the square to a rectangle $2a$ by $2b$, and the circle to an ellipse with major and minor axes a and b. Then we see that the rectangle becomes $2a$ by $2b$. So, the area of the rectangle will be $4ab$. If the same ratio as for a circle holds, then the area of an ellipse should be πab..

If you want to prove this by overlaying two models, you can print an ellipse in OpenSCAD by stretching a cylinder. Let's print an ellipse with a semimajor axis of 40 mm and a semiminor axis of 20 mm, just 5 mm tall. We can do that with this one-line OpenSCAD model:

```
scale([2, 1, 1]) cylinder(h = 5, r = 10);
```

Then you can print a rectangular solid 5 mm high by 40mm long by 20mm wide with this line of OpenSCAD:

```
cube([40, 20, 5]);
```

When you overlay them (Figure 9-32), you can see that indeed the ellipse will be inscribed in a consistently stretched rectangle. We can also make the same scaling arguments we did to find the equation of an ellipse to get equation for the area, which is indeed πab.

FIGURE 9-32

An ellipse inscribed in a rectangle

Circumference of a Circle and an Ellipse

The very definition of π is the relationship between a circle's diameter and the distance around its outside (its circumference) and we know that the circumference of a circle is $2\pi r$. An ellipse, however, is another matter. When you draw an ellipse (or use the 3D printed model) you can also find the circumference, the equivalent distance all the way around it.

Lay down a piece of string around the ellipse, mark off the circumference on the string, and then straighten out the string on a ruler to measure the circumference. Likewise, measure a and b. Do you see any obvious relationship?

Bizarrely, there is no simple equation for the circumference of an ellipse, for reasons that require calculus to compute. Around 1914, the Indian mathematician Srinivasa Ramanujan came up with this approximation (a squiggly line on top of an equals sign means "approximately", also sometimes shown as two squiggly lines) for the circumference of an ellipse (C):

$$C \cong \pi[3(a + b)] - \sqrt{(3a + b)(a + 3b)}$$

How close was your measurement of the string to what you can compute, given your measurements of a and b? It is a tribute to how hard this problem is that this approximation is only about 100 years old. Ramanujan, who lived in British India during its colonial period, did not have any formal training in higher mathematics but came up with this and other innovations largely on his own. He also developed better, more complex approximations as well.

Search on his name and "ellipse circumference" (sometimes also called "ellipse perimeter") for more. Unfortunately, he died at age 32, or who knows how many other simplifications he would have discovered. The full solution involves an advanced calculus equation called an "elliptic integral", beyond even our *Make: Calculus* book.

Ellipses in the Wild

Ellipses arise in many real-world applications, from modeling the orbits of planets (look up "Kepler's Laws") to the shape of a (nearly circular) park called "The Ellipse" near the White House in Washington, DC. One characteristic of physical elliptical walls is their ability to focus sound or other waves. To understand this, we need to pull in a few things from Chapter 8, as well as what we learned in this chapter.

Imagine that we had an elliptical smooth-walled, shallow pool, and we put a little piston in the water at one focus that moved up and down in a sine wave. What would happen? As each wave front went forward, it would eventually hit the wall. As we saw in Chapter 8, a reflected wave comes back out at the same angle it came in but reflected in the opposite direction. In the case of a curved surface, we can think of it reflecting off a tiny flat wall that just touches the curve (called a "tangent" line). If we look back at Figure 9-26, we can think of the edge of the slider as being that tangent line, and the line of the string as the line of the reflected wave in the ellipse.

Now, if we think about the fact that any line from a focus to a point on an ellipse, and back to a focus is the same distance, this means that all those points stay in phase with each other. In fact, all the reflections from a source at one focus will be focused at the other one (hence the name).

Modeling this phenomenon accurately with waves bouncing around an elliptical chamber is a bit beyond the simple wave models we used in Chapter 8; a static snapshot of waves interfering is a bit hard to interpret anyway. Mathematical physicist Nils Berglund has many amazing simulations on his YouTube channel (details in References at end of this chapter).

There are physical manifestations of focusing. Some medical ultrasound devices to break up gallstones and kidney stones have an emitter at one focus with a half-elliptical reflector, and the patient's organ needing treatment at the other focus. The process, called *lithotripsy*, focuses enough energy to break up the stones without surgery.

Somewhat more pleasant is the whispering gallery phenomenon. In some buildings with curved walls (famously St. Paul's Cathedral in London, and a particular place in Grand Central Station in New York) people can talk softly near one focus and be heard perfectly well at the other focus across a large open space. It is tricky to make a whispering gallery, even on purpose, so they are not very common. In the next chapter we talk about parabolic reflectors, which are a little more practical for real-world amplification applications.

Terminology and Symbols

Here are some terms and symbols from the chapter you can look up for more in-depth information:

- analytic geometry
- circle
- conic sections
- eccentricity
- ellipse
- focus (plural foci)
- hyperbola
- major and minor axes
- parabola
- rotation
- slant angle
- slant height
- translation

Chapter Key Points

In this chapter, we reviewed the definitions and basic geometry of conic sections, then zoomed in on the circle and ellipse. We discovered how to draw an ellipse with string based on the definition of an ellipse as a set of points such that the sum of the distances from any point to the two foci is constant. Then we moved on to a 3D printed model (also using string to demonstrate key properties). We then used these geometry-based models to move into analytic geometry, deriving equations for the ellipse and its foci in Cartesian coordinates, as well as how to translate and rotate an ellipse. Finally, we discussed places where ellipses occur naturally, and the physical implications of these shapes.

References

Here are some sources to go into more depth on the topics in this chapter:

Merzbach, U. C., and Boyer, C. B., (2011) *A History of Mathematics* (3rd ed.). John Wiley and Sons. This book is a broad resource on the worldwide history of mathematics. It is written assuming that the reader knows mathematical terminology at the college level but is a very comprehensive guide to use as a reference.

Mathematical physicist Nils Berglund's visualizations can be found on **YouTube** (https://www.youtube.com/@NilsBerglund). Search the videos with the keyword "ellipses" to see phenomena described in this chapter. He also has open sourced the code for these on Github; he has details of his resources in the "About" section on his YouTube channel.

10 Parabolas and the Quadratic Formula

3D Printable models used in this chapter

See Chapter 2 for directions on where and how to download these models.

- `parabola_slider.scad`
  - This model creates a parabola given various of its dimensions
- `conic_parabolas.scad`
  - This model creates a cone with many parabolas sliced into it
- You will also need
  - Elastic cord. We used 0.8 mm elastic cord, also called beading thread or crafting cord. You will need about 1 foot per model you choose to make.
  - A ruler or calipers
  - A protractor
  - Some graph paper, or the Cartesian x-y plane model from Chapter 5
  - A pipe cleaner (sometimes called a chenille stem)

Parabolas

In Chapter 9, we introduced all the conic sections (circle, ellipse, parabola, hyperbola), and went into depth on the ellipse and circle. In this chapter, we explore the parabola — the curve that is formed when we slice across a cone with a cut parallel to the side of the cone. Any shallower cut gives us an ellipse (or circle); any steeper, and we get a hyperbola.

We can think of a hyperbola as sort of an ellipse flipped inside-out (more on this in Chapter 11) and the parabola as the transition between these two. A parabola has some interesting properties, and they appear frequently in practical applications, including mirrors that focus light or sound (more on that later in the chapter).

For any given cone, only one cut *angle* produces a parabola (or a circle). However, it is possible to cut at that angle in multiple places on the cone, creating parallel cuts and thus different parabolas. It is analogous to cutting different circles from the same cone.

We can see this by using the model `conic_parabolas.scad`. This model creates multiple parallel cuts across a cone (Figure 10-1), showing a variety of parabolas that can be cut from the same cone (Figure 10-2) while keeping the cut angle the same. There are small brims on each cut to make it obvious how the parts go together and to make the model stand up better. You might still need a bit of museum tack between the layers to make the model hold together. (We turned the slices in Figure 10-2 upside-down and rotated a bit from their orientation in Figure 10-1's cone to make the slices a little easier to see.)

FIGURE 10-1

Parabola slices on the cone

FIGURE 10-2

Resulting parabolas

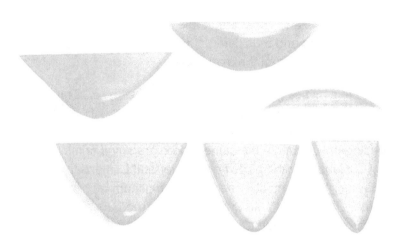

The model has several parameters you can change:

- h = height, in mm, defaulted to 50 mm
- r = radius, which is the amount of brim on each section, defaulted to be the same as the height.
- cuts = horizontal offsets between the various cuts, in mm.

In this chapter, we assume that you know how to multiply two polynomials together, like $(x + 2)(x + 3)$, and have some general ideas on how to factor the result of such a multiplication. If not, start at the FOIL sidebar later in the chapter, and look up "multiplying polynomials" and "factoring polynomials" for more before you go on.

Focus and Directrix

In addition to its conic section definition (Chapter 9), a parabola is also defined as the curve that maintains an equal distance from a point (the *focus*) and a line (called the *directrix*). Unlike an ellipse, this distance is not the same all the way along the curve. A parabola is shaped sort of like the letter U, and some are stretched out, or made narrower (closer to a V but not pointy where it changes direction).

To create a parabola, we would first draw a point (the focus) and then the line (directrix) as in Figure 10-3. The vertex of the parabola is halfway between the focus and directrix. Then, we want to find more points halfway between the two, moving away from the vertex in either direction.

Remember that lines that have one hash mark are equal in length to each other, lines with two hash marks are equal to each other, and so on.

Ellipses and hyperbolas can be defined in terms of a focus and directrix as well. We will talk about this way of thinking about an ellipse in Chapter 11, when we compare it to a hyperbola and explore the idea we just noted of a hyperbola as an inside-out ellipse. Since a parabola is the transition case between an ellipse and a hyperbola, we can think of a parabola as an ellipse with one focus fixed and the other at infinity.

Drawing a Parabola

As we just mentioned, a parabola can also be described as a curve that has every point an equal distance from a point (the focus) and the shortest distance to a line (the directrix). In Figure 10-3, we show these distances marked with little hash marks. Lines with one hash mark are equal in length to each other, the ones with two are equal, and so on. The directrix is shown in red.

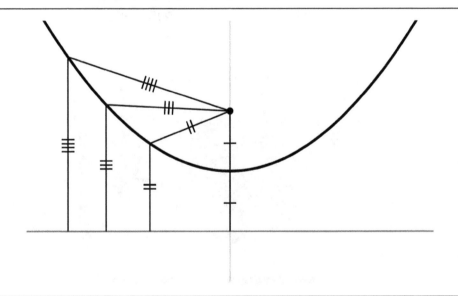

FIGURE 10-3

A parabola, showing its directrix line (red), focus, and lines of equal distance

The Parabola Model

Our model `parabola_slider.scad` allows you to slide along the parabola and see how these two distances stay the same as each other (Figure

10-4). Unlike the ellipse, the sum of these distances is not a constant. Model parameters you may want to change (all dimensions in mm) are

- `length = 100;`
  - longest dimension of model
- `a = length/4;`
  - focus to vertex distance — does not have to have this relationship to length
- `holesize = 5;`
  - size of the hole in slider and at focus
- `h = 8;`
  - thickness of model
- `wall = 2;`
  - width of directrix frame

For practical reasons (wanting to keep the string a constant length), this model shows the two equal distances a little indirectly. The model incorporates the focus, the directrix, and a line parallel to the directrix. We will get to how to use it after we describe how to put it together.

FIGURE 10-4

Parabola 3D printed model

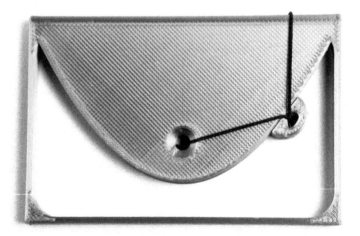

Note that you do not want to make the model too small, since the slider will become challenging to print accurately if you go much below the default dimension values here. In fact, you probably want to make this model as large as your 3D printer will allow, since that will make the measurements we suggest later easier.

Assembling the Model

Like the ellipse model in Chapter 9, some assembly of this model is required. Take a length of slightly stretchy cord about twice as long as the longer dimension of the model (Figure 10-5). Technically a non-stretchy cord would be more accurate, but it is very hard to tie off and move the cord if there is no give to it. Run the cord from the back of the model to the front through the focus, then through the slider from front to back (Figure 10-6). Loop the cord over the rounded top of the model then through the slider again from front to back (Figure 10-7). Finally, tie off the cord on the back side of the model (Figure 10-8). Figure 10-9 shows the completed model.

FIGURE 10-5

Initial string placement in the model

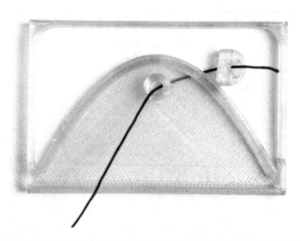

FIGURE 10-6

*Continuing around
the back side of the
model*

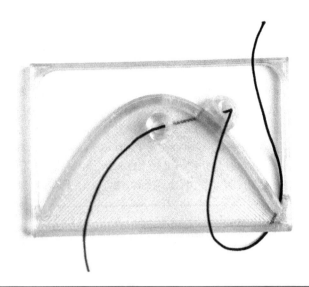

FIGURE 10-7

*Loop the cord over
the rounded top of
the model then
through the slider
again from front to
back*

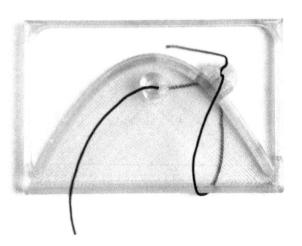

FIGURE 10-8

Tie off the string

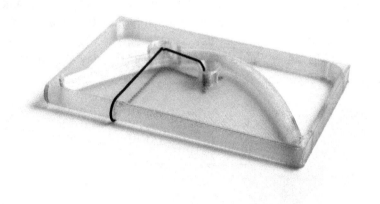

FIGURE 10-9

The completed model

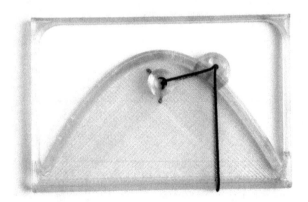

Learning with the Model

Once you have assembled the model, adjust it so that the cord makes a right angle with the rounded edge. It should want to be at this angle, since the string must stretch for it to reach other angles. If the friction of the model surface is high or it catches on a print artifact, it may be slightly off. Just to make it simpler to describe, we will call the rounded edge the top of the model (as oriented in Figure 10-10) although a parabola can be oriented in any direction.

Now you can measure the distance from the center of the focus hole to the curve with a ruler (along the string, as in Figure 10-10) and the distance from the curve straight down to the directrix (use the vertical string as a guide, as in Figure 10-11). You will find they are equal.

FIGURE 10-10

Measuring from the focus to the slider

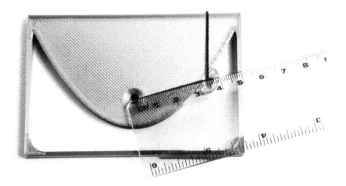

FIGURE 10-11

Measuring from the slider to the directrix

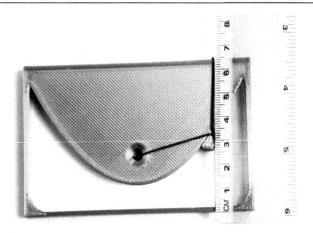

Next, consider the vertical distance from the top of the model to the slider. The sum of that distance and the distance to the directrix is a constant, since the bottom of the model (representing the directrix) and the top are parallel lines. Therefore, the length of the string stays equal as you move around the model.

Now, move the slider around the parabola a bit farther, and convince yourself that these relationships hold there, too. Be careful that the string makes a right angle with the top of the model.

Consider what happens if a parabola has its open side pointed up, or down, or sideways, or at some weird angle. (You can do this by just turning this model appropriately.) Where will the directrix be in relation to the focus? (Answer: If the open side of the parabola points down, then the directrix will be above the parabola, and similarly for right and left opening parabolas.)

FOIL

If we want to multiply together two polynomials, like $(x + 2)(x + 3)$, there is a simple way to remember how to do it that has the acronym FOIL, for "first, outer, inner, last." To multiply these two expressions times each other, we multiply term by term. We multiply x times x (first terms), then 2 times x (inner), then x times 3 (outer), then finally 2 times 3 (last). The answer we would get here is

$$(x + 2)(x + 3) = x^2 + 2x + 3x + 6 = x^2 + 5x + 6$$

If this is new to you, you might want to look up "factoring polynomials" as well to see the reverse process and get some tricks and tips.

Equation of a Parabola

Now that we have some definitions of a parabola in terms of its focus, directrix and vertex, is there an equation we can write that describes a parabola in terms of x and y coordinates? Let's say we draw a parabola so that its vertex is at $x = 0$ and $y = 0$. We will say that its focus is at $x = 0$ and $y = f$, and that the directrix is the horizontal line $y = -f$.

We need to find the shortest distances from the focus to the parabola and the parabola to the directrix and set them equal to each other to get our equation. This is shown in Figure 10-12, and you can also note the dimensions on the 3D printed model.

FIGURE 10-12

Anatomy of a parabola

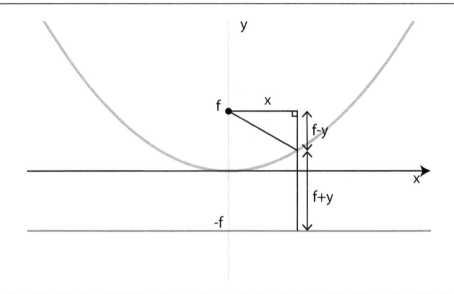

First, let's figure out the distance from the directrix to a given point on the parabola. If we want to find the distance to some arbitrary point (x, y), we can use the Pythagorean Theorem to make ourselves a little right triangle, the hypotenuse of which is the dimension we want. We know the distance from the focus to the vertex is f, by the definition of f.

The distance from a given point on the parabola to the focus in the y direction is $f - y$, since the point we are looking at is lower than the focus. In the x direction, the distance from a given point on the parabola is just x, since we are assuming that the parabola is centered on the y axis. Therefore, the distance (d) from the focus to the parabola, using the Pythagorean Theorem, is

$$d = \sqrt{x^2 + (f - y)^2}$$

We can see that the shortest distance from a given point on the parabola to the directrix will just be a line straight down. Since the directrix is distance a below the x axis, this distance (d) is

$$d = f + y$$

From the definition of a parabola, we know that the distance from focus to parabola equals the distance from parabola to directrix, so making these two values of d equal to each other, we get

$$\sqrt{x^2 + (f - y)^2} = f + y$$

We can square both sides to get

$$x^2 + (f - y)^2 = (f + y)^2$$

When we multiply all this out, we get

$$x^2 + f^2 - 2fy + y^2 = f^2 + 2fy + y^2$$

A lot of these terms cancel out, and we are left with

$$x^2 = 4fy$$

Or, if you prefer,

$$y = \frac{x^2}{4f}$$

If f is a positive number, then the open part of the parabola is in the upward direction. If f is a negative number, the parabola is like an upside-down U. If the parabola is not lined up with the axes like we have shown here, the algebra gets messier, but the principle is the same. If instead we have the curve,

$$x = \frac{y^2}{4f}$$

then if f is positive, the parabola will open to the right. If f is negative, the parabola will open to the left. Note that to keep the algebra simple, we assumed the vertex of the parabola is at the point $(x, y) = (0, 0)$. We can always shift a parabola right or left or up and down by subtracting an offset from x or y.

Mark x-y axes on a sheet of paper or take the Cartesian x-y plane we created in Chapter 4. Take a pipe cleaner and bend it into a parabola shape (Figure 10-13). Then test yourself to see on the axes how these parabolas are oriented.

$$y = \frac{x^2}{4f}$$

$$y = \frac{-x^2}{4f}$$

$$x = \frac{y^2}{4f}$$

$$x = \frac{-y^2}{4f}$$

FIGURE 10-13

*Parabola orientation
test*

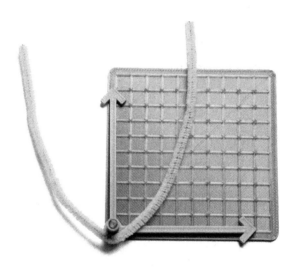

The answers are:

$y = \frac{x^2}{4f}$ (opens upward, directrix $y = $ -f)

$y = \frac{-x^2}{4f}$ (opens downward, directrix $y = f$)

$x = \frac{y^2}{4f}$ (opens right, directrix $x = $ -f)

$x = \frac{-y^2}{4f}$ (opens left, directrix $x = f$)

Pick a value for f. What would be the equation of the directrix in each case? Remember that the vertex will be halfway between the focus and the directrix, which are both a distance equal to f from the curve at that point.

Translating the Parabola

To this point we have been talking about parabolas with a vertex at x and y both equal to 0. If we want to raise the parabola above the x axis, or shift it left or right from the y axis, we would subtract the offset in y or x respectively. If we have a parabola with a vertex at $(x, y) = (2, -3)$ and a focus at $(2, -2)$, we see that the vertex and focus are both on the line $x = 2$. The vertex is 1 away from the focus, so $f = 1$. The vertex is below the focus, so the parabola opens upward. The focus and vertex give us the equation for the parabola:

$$y - (-3) = \frac{(x-2)^2}{4 * 1}$$

or, multiplying this out,

$$y + 3 = \frac{1}{4}\left(x^2 - 4x + 4\right)$$

$$y = \frac{1}{4}x^2 - x + 1 - 3 = \frac{1}{4}x^2 - x - 2$$

Thus, the equation for our parabola with vertex (2, -3) and focus (2, -2) is

$$y = \frac{1}{4}x^2 - x - 2$$

As we will see, this is a very common and convenient way to write a parabola. In general, one standard way to write it is

$$y = ax^2 + bx + c$$

where a, b, and c are constants (called the *coefficients* of their respective powers of x). In the case we just talked about, $a = \frac{1}{4}$, $b = $ -1, and $c = $ -2.

Equations like this that have terms that are positive integer powers of a variable are called *polynomials*. A polynomial with a square as its highest power would be called a "polynomial of degree 2," more commonly known as a *quadratic equation*. We will see quadratic equations in other forms later, but they are still called that if there is a squared term (and no higher-power ones).

The points (if any) where the parabola crosses the x axis (when $y = 0$) are called the *roots* of the polynomial. In the next section we will talk about how to find them using a technique called *completing the square*. The word *quadratic* comes from a Latin phrase meaning to make something square, so we will see how all this fits together.

Quadratic Equations

We know how to solve an equation like this:

$$x^2 = 4$$

The answer is that $x = 2$ and -2, since both of those, multiplied by themselves, equal 4. An equation with a square in it will always have two solutions. Now, what happens if we instead have

$$y = x^2 - 4$$

If we graph this, we will get a parabola which faces upwards, and which crosses the x axis at positive and negative 2. That is because $y = 0$ on the x axis, so

$$0 = x^2 - 4$$

Which is just the same as the equation at the beginning of this section. Thus, the roots of this parabola are positive and negative 2. Since these are symmetrical around the y axis (where $x = 0$) the vertex must be at $x = 0$. When we substitute $x = 0$ in

$$y = x^2 - 4$$

we get $y = -4$, and the vertex is at (0, -4). Note that there is no plus and minus here, since y is not squared. What about the directrix and focus? We saw earlier that to find the y value of the focus f for a parabola with its vertex at (0, 0) we could use the formula

$$y = \frac{x^2}{4f}$$

and in this case, we have

$$y = x^2 - 4$$

This says that we shift the focus we would have gotten without the "-4" down by 4. We want $4f$ to be equal to 1 to get our equation. Thus, the focus is at $x = 0$ and $y = -4 + \frac{1}{4} = -3.75$ (Figure 10-14).

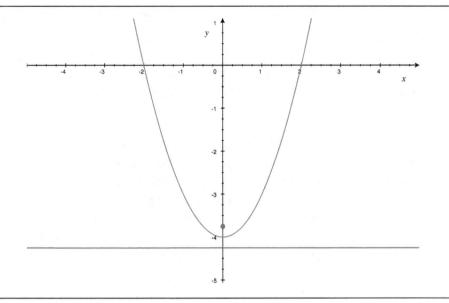

FIGURE 10-14

Parabola, its roots, focus (red dot) and directrix (red line)

Roots: The Fundamental Theorem of Algebra

In earlier encounters with algebra, you probably learned to solve equations like

$$0 = 5x - 10$$

To solve it, you would shuffle around the terms to get the x by itself, first by adding 10 to both sides of the equation:

$$5x = 10$$

and then dividing by 5.

$$x = \frac{10}{5} = 2$$

We can refer to 2 as the *root* of this equation — the value of x that makes it a true statement. But what can we do if we do not have a good way to get x by itself? For example, what can we do if we have an equation that is a function of x, but we have terms in x^2, x^3 or some mix of terms of various (integer) powers?

The *Fundamental Theorem of Algebra* says that if we have a polynomial with terms that are integer powers of x, the polynomial will have a number of roots equal to the highest power of x. That is, if we have an equation in x^3, it has three roots; quadratic equations have two. This is sometimes called *d'Alembert's Theorem*, for Jean le Rond d'Alembert, a French mathematician, physicist, and contributor to music theory who was active in the mid-1700s. (Technically d'Alembert just did sort of a first draft of the theorem, which was cleaned up by German mathematician and physicist Carl Frederich Gauss in 1799, and further adapted well into the twentieth century.)

The core of the Fundamental Theorem of Algebra is that if we have a polynomial in powers of x, it can be *factored* into a series of terms that can be multiplied together to get the original back. Some polynomials can be factored manually, with some mix of luck and skill. (Look up "factoring polynomials" if you have not seen the technique.) In the next section we learn a technique for quadratic equations, so we do not have to bother with this guesswork. But let's say we have the polynomial

$$x^2 - 6x - 8 = 0$$

which we can factor into $(x - 2)(x - 4)$. (Multiply it out with FOIL to check.)

$$(x - 2)(x - 4) = 0$$

Now, the only way that this product can be zero is if one (or both) of the terms is zero. So, we get $x - 2 = 0$ (or $x = 2$), and $x - 4 = 0$ (or $x = 4$).

The roots of this equation are 2 and 4. In principle one can do this process for any polynomial, but beyond quadratic equations things get messy. Note that there is no y variable here. We are talking about finding a pair of points, NOT graphing a curve of y versus x. We will get back to what this has to do with parabolas in a little bit.

There are also special cases. For example, roots can be repeated, as in the equation

$$x^2 - 4x + 4 = 0$$

which factors into

$$(x - 2)(x - 2) = 0$$

has duplicate roots 2 and 2. This is not considered to violate the theorem.

Quadratic equations come up all the time in physical problems, in circumstances where we do not really care about graphing a parabola of it for all possible values of x. In the case of finding the roots of a quadratic, we are interested in just a (pair of) answers, not the curve. Say for example I want to lay out a room in a house that is three feet longer in one dimension than another, and a total area of 120 square feet. The length of one side is $x + 3$, the length of the other is x. The area would be

$$x(x + 3) = 120$$

Multiplying out and taking the 10 to the other side, we can get it in our solvable form,

$$0 = x^2 + 3x - 120$$

When we use the *quadratic formula* (derived in the next sections of this chapter) to solve this equation (it does not factor neatly) we discover there are two answers: 9.55 feet and -12.55 feet. Obviously, there is no way to make a negative-area room, so our answer is that our room will need to be 9.55 by 3 + 9.55, which we can see gets us to 120 (with a bit of rounding error).

Physical problems described by a quadratic often have one nonsense answer (called a *spurious solution,* or extraneous solution) and one useful one, or there might be two useful ones and either would serve your purpose. As we discuss in the next section on completing the square, sometimes roots contain *imaginary numbers,* involving the square root of -1, and that may mean there are no real solutions to your problem.

So, what does this have to do with parabolas? If we graph a quadratic function, we can read the roots off the graph where $y = 0$. Were you to graph our example, you would see it crosses the x axis at $x = $ -12.55 and $x = 9.55$, the roots we just found. (Test this example out after you learn the quadratic formula in the next two sections.) To put it another way, when we graph our parabola for values of x other than the roots, those values of x do not solve our problem. If $x = 1$, $x + 3 = 4$, and our room would be 400 square feet, not the 120 we wanted.

What happens beyond quadratics, for higher powers of x? If you need more, look up "cubic equation formula" for equations that go up to x^3 and "quartic equation" for equations involving x^4. Linear algebra techniques are needed to work through these more-complex systems in an orderly way. (The mathematicians Neils Abel and Évariste Galois proved that in general no exact solution exists for equations involving x^5 , except for some special cases.)

This goes well beyond our scope here, but if you need to know how to solve these higher-order equations, start by searching on *Vieta's Formulas*. These are named after François Viète, who lived in France in the mid-1500s and is also known by the Latin version of his name, Franciscus Vieta. We met Viète, who was quite the polymath, back in Chapter 6, when we talked about prosthaphaeresis and its foreshadowing of logarithms. Like Newton would about 100 years later, he also had to invent precursors of some of the notation we take for granted today to express polynomials.

The bottom line, though, is that these techniques let you solve for x beyond the familiar environs of a linear equation, opening up abilities to figure out numerical answers to whole classes of new problems. Next, let's see where we got our answer to our little 120-square-foot area problem.

Completing the Square

Earlier in the chapter we looked at a parabola that was centered at $x = 0$. What happens in the more general case, where it might be shifted over both up and left or right? Earlier in this chapter we saw that the equation for this general case is

$$y = ax^2 + bx + c$$

and we established in the last section about the Fundamental Theorem of Algebra that this equation will have two roots, which we can find when $y = 0$. Before we go too far, though, why might we want to find these roots of a quadratic, other than in a math quiz? As it turns out (and we will see later in the chapter) quadratic equations come up all over the place in physical situations. We need to be able to find the roots to get answers to these practical problems.

A technique called *completing the square* is a very old graphical method of finding these roots. Babylonian texts 4000 years old discuss a variant, according to Merzbach and Boyer's *History of Mathematics* that we have cited in earlier chapters. Apparently, they invented techniques for solving these equations even before they invented an alphabet, since quadratics come up often in practical calculations of area and volume, as we saw in the last section. The Babylonians were not an anomaly, though: Arab mathematician Al-Khwarizmi (who lived about 1300 years ago) also documented its use in an algebra treatise.

We saw in an earlier section how to graph a parabola if we did not have that pesky bx term. Completing the square sort of gets rid of that term with some rearrangement. Suppose we first took our c out of the mix on the left side.

$$y - c = ax^2 + bx$$

We are trying to find roots first, which means we are going to set y = 0:

$$0 - c = ax^2 + bx$$

It would be very handy if the left-hand side was (x + something) squared. Let's divide by the constant, a, and see what we have left.

$$\frac{-c}{a} = x^2 + \frac{b}{a}x$$

Now, what would give us that middle term if we want (x + something) squared? (If you do not know how to square an expression, see the "FOIL" sidebar.) To make it a little easier to see, we can leave this algebra for a minute and use a simpler example to remind ourselves how to square an expression like (x + p):

$$(x + p)^2 = (x + p)(x + p) = x^2 + px + px + p^2 = x^2 + 2px + p^2$$

Note that we get two px terms. The middle term is twice the value of the "something" we add to x (p, in this case), and in the final equation we add a constant of our "something" squared (Figure 10-15).

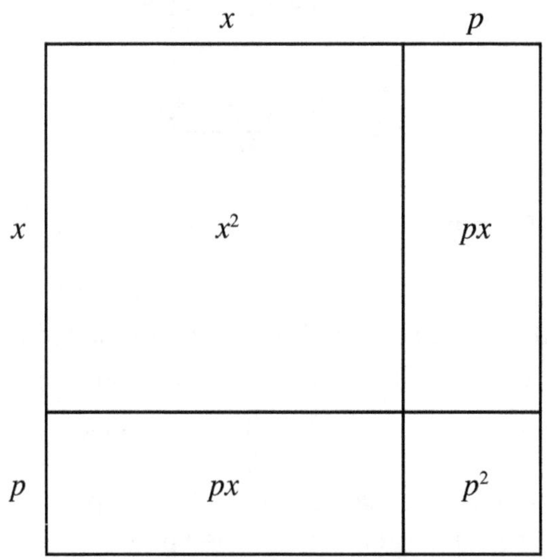

FIGURE 10-15

Completing the square

If we go back to our general equation, and add the square of $\frac{b}{2a}$ to both sides, we get this:

$$\frac{-c}{a} + \left(\frac{b}{2a}\right)^2 = x^2 + \frac{b}{a}x + \left(\frac{b}{2a}\right)^2$$

Now, despite your possible feeling of rising panic, this really IS simpler, because we can tidy up this mess to be

$$\frac{-c}{a} + \frac{b^2}{2a^2} = \left(x + \frac{b}{2a}\right)^2$$

Remember that we are squaring half of $\frac{b}{a}$, which is where the 4 on the bottom comes from. And voila, now there is no term in just x. Let's make the left-hand side a little neater by putting everything over a common denominator of $4a^2$.

$$\frac{b^2 - 4ac}{4a^2} = \left(x + \frac{b}{2a}\right)^2$$

Taking the square root of both sides, we get

$$\pm\frac{\sqrt{b^2 - 4ac}}{2a} = x + \frac{b}{2a}$$

where "±" means we get a positive and negative square root, just as we did in the earlier example. Solving for x we get the two roots:

$$x = \frac{-b + \sqrt{b^2 - 4ac}}{2a}$$

$$x = \frac{-b - \sqrt{b^2 - 4ac}}{2a}$$

The Quadratic Formula

This equation is called the *quadratic formula*. The term under the square root is called the *discriminant*, and tells us how many roots the equation has. If $b^2 - 4ac = 0$, then the equation only has one root, $\frac{b}{2a}$. If $b^2 - 4ac$ is greater than zero, then the equation has two roots.

But what if $b^2 - 4ac$ is less than zero? That would be asking us to take the square root of a negative number. When people solved this by drawing actual squares, this would have asked them to draw a square with a negative area, which obviously is not possible. Hence, these were called *imaginary* solutions (as opposed to "real" solutions).

Graphically, having imaginary solutions means that the parabola does not cross the x axis. For example, $y = x^2 - 2x + 2$ looks innocuous enough, but its discriminant is $4 - 4 * 1 * 2 = -4$. Graphing this parabola shows that it indeed does not cross the x axis (Figure 10-16). In our *Make: Calculus* book we explore imaginary numbers, but for now you can just think of these solutions as showing us that the parabola does not cross the axis.

FIGURE 10-16

A parabola with no real roots

As an exercise, try finding the roots of the equation we came up with back in the Fundamental Theorem section:

$$0 = x^2 + 3x - 120$$

Here, $a = 1$, $b = 3$, and $c = -120$. Prove to yourself that the roots are indeed -12.55 and 9.55 by using the quadratic formula.

Focus, Directrix and Vertex

Originally, we had an equation like this to find our focus:

$$y = \frac{x^2}{4f}$$

The vertex of this parabola is at (0, 0), the focus is at (0, f) and the directrix is the line $y = -f$. Can we shuffle around our general parabola equation so that we are just shifting our coordinates in x and y and can read off our vertex, focus, and directrix? If we can change our coordinates to x minus an offset and y minus a (different) offset, but otherwise having the form above, we can do that.

Let's go back to our process of completing the square, but now we are not making y = 0. This time, we just leave the y alone, since we want a relationship between x and y.

$$y - c = ax^2 + bx$$

Complete the square as before:

$$\frac{y - c}{a} + \frac{b^2}{4a^2} = \left(x + \frac{b}{2a}\right)^2$$

Multiply through by a:

$$y - c + \frac{b^2}{4a} = a\left(x + \frac{b}{2a}\right)^2$$

We can think of this as shifting our x coordinate system by $\frac{-b}{2a}$, and our y coordinate system by $\frac{-b^2}{4a} + c$. Why the negative? If we start off shifted, we will want to think of subtracting off these offsets to get back to our standard form.

Therefore, the vertex will be at

$$(x, y) = \left(\frac{-b}{2a}, c - \frac{b^2}{4a}\right)$$

The focus will be offset by $\frac{1}{4a}$ in y, and will be at the coordinates

$$(x, y) = \left(\frac{-b}{2a}, c - \frac{b^2}{4a} + \frac{1}{4a}\right)$$

or, in a form a little easier to read

$$(x, y) = \left(\frac{-b}{2a}, \frac{4ac - b^2 + 1}{4a}\right)$$

Note that the sign of the coefficient a takes care of whether the focus is above the vertex (for positive a) or below it (for negative a, and a downward-opening parabola). The directrix will be the line

$$y = c - \frac{b^2}{4a} - \frac{1}{4a} = \frac{4ac - b^2 - 1}{4a}$$

Armed with this, we can see how varying a, b, and c will change the position and shape of the graph of our parabola. We walk through that in the next section.

If y is the squared variable, then we would need to go through an equivalent exercise in completing the square of the y variable. As a side note, some authors will write the general equation of a parabola as

$$(y - k) = \frac{(x - h)^2}{4p}$$

or perhaps

$$4p(y - k) = (x - h)^2$$

We prefer the way we have done it. It lets us think of starting with one idealized parabola with its vertex at (0, 0) which we shift and expand based on the coordinate system that is convenient for whatever we are trying to do.

Changing the Curve

In the last section, we saw that the location of the vertex of our parabola was a function of a, b, and c. Now, assuming that we are talking about the x variable being the squared variable (and therefore that the parabola opens up or down as opposed to left and right), let's see what this variation looks like.

The contribution of the coefficient c is easy to understand. It just moves the whole parabola up and down (along with the focus and directrix) as we see in Figure 10-17. The black arrow shows the direction of motion of the vertex of a parabola as the coefficient c goes from negative numbers (like the red parabola) through zero (green parabola, touching the x axis) to positive numbers (blue parabola).

FIGURE 10-17

Parabolas as the coefficient c is varied

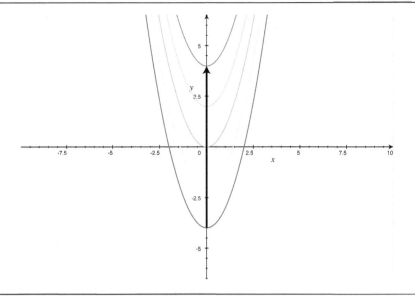

The coefficient a is a scaling factor that primarily makes the parabola wider or narrower. Our experiment cutting different parabolas from the same cone back in Figures 10-1 and 10-2 could be thought of as an exercise mostly in changing a, at somewhat arbitrary intervals. If we think of the cone in Figure 10-1 as being infinitely tall, we can think of the slices near the bottom as zooming in on the parabola more tightly.

Now, we know the distance from the vertex to the focus and from the vertex to the directrix are both $\frac{1}{4a}$. That means as the coefficient a gets bigger, that distance gets smaller. As that distance gets smaller, the parabola widens out to maintain the same distance from the focus to any given point on the parabola and the (also now closer) directrix. Thus, the width is a function only of the parameter a (Figure 10-18). That means that no matter what the value of the coefficients b and c are, the parabola has to be the same shape — just shifted around. The arrows in Figure 10-18 show how the parabola changes as we go toward smaller negative values of a (blue parabolas), then through $a = 0$ (purple line, on the x axis) and then toward increasing positive values of a (red parabolas).

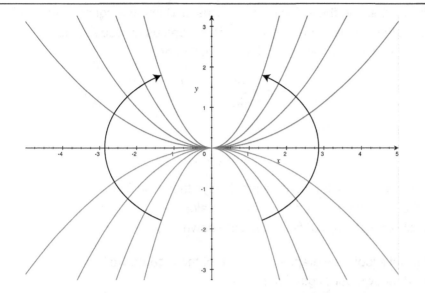

FIGURE 10-18

Parabolas as the coefficient a is varied

How does the b coefficient affect where the parabola is placed? As we saw in the last section, the vertex of the parabola moves around with the x and y coordinates functions of a, b, and c:

$$x(a, b) = \frac{-b}{2a}$$

Note that there is no dependence on c.

$$y(a, b, c) = c - \frac{b}{4a}$$

Let's say we hold a and c constant, and just vary b. As b increases, the x coordinate of the vertex shifts to the left (since there is a negative sign on the b). The y coordinate, though, has a dependence on b^2.

If we were to use these equations for x and y vertex positions, hold a and c constant, and vary b, we could plot the resulting points on an x-y graph. It is still kind of hard to see what is going on, though. Maybe we can get our b-dependence back into a function of x and y again to see what curve we get when b varies. To do that, solve for b as a function of x first, for the coefficient a held constant:

$$b = -2ax$$

Then substitute this value of b (in terms of x) into the equation for y, with coefficients a and c held constant. This gives us an equation for y as a function of x again, instead of as a function of b.

$$y = c - \frac{b^2}{4a} = c - \frac{(-2ax)^2}{4a}$$

or

$$y = c - ax^2$$

Now, that is an interesting result. It says that if we fix our values of a and c, the vertex of the parabola will slide along a curve that is the original parabola when $b = 0$, flipped upside down about its vertex.

Now let's look at a graphical version of the algebra we just did. Figure 10-19 shows our original parabola,

$$y = a^2 + bx + c$$

for which b did not equal zero. We know that because the vertex is at $x = \frac{-b}{2a}$, which means the vertex has to be at $x = 0$ if $b = 0$.

FIGURE 10-19

The original parabola

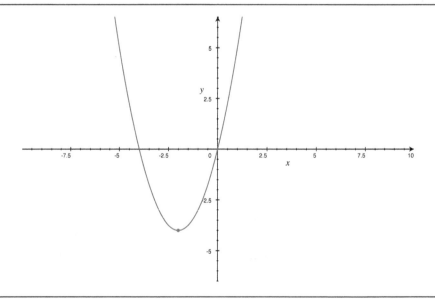

Next, we want to find the parabola $y = ax^2 + c$ (in other words, setting $b = 0$ while leaving a and c alone). This is the pale green, opening-upward parabola in Figure 10-20. Note that it is still the same shape as the original parabola but shifted.

Then we need to draw the parabola for the opposite sign of the coefficient a. (In other words, draw $y = c - ax^2$.) This is the black downward-facing parabola in Figure 10-20, a result of flipping the parabola with b = 0 over in the y coordinate. This black parabola, as we established, is the path the vertices of parabolas follow if we hold the coefficients a and c constant, but vary b. The arrow notes the direction the vertices travel as b goes from negative to positive values.

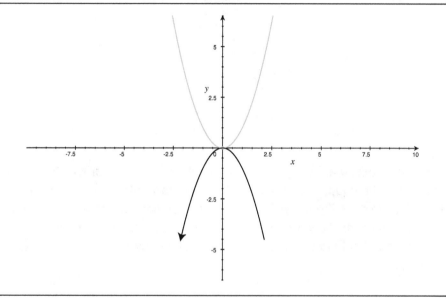

FIGURE 10-20

Parabola shifted to the b = 0 point, and the reflection with –a

Figure 10-21 shows an assortment of parabolas, all of which have the same values of a and but are varying b. The vertices of all these parabolas lie on the black parabola opening the opposite way, and you can think of it as a track that parabolas follow as b changes from a negative number (red parabola) to positive (purple and blue parabolas). Notice that all of them have the same shape as each other, but their vertices slide along the black "track" as b changes.

FIGURE 10-21

Parabolas as the coefficient b is varied

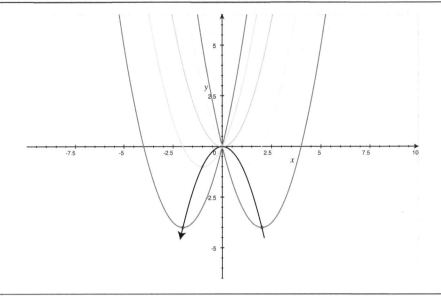

If we were talking about right- or left-opening parabolas, for which the y-variable is the squared one, all these variations would happen 90° from the ones we describe here. As a final note, rotating parabolas is messier algebraically than we want to get in this book, but the same general principles will hold regardless of what coordinate system you choose to view them.

Parabolas in the Wild

To this point, the math in this chapter might seem rather abstract, and you might wonder why parabolas deserve so much attention and analysis. However, parabolas come up in the modeling of many practical (and very useful) situations.

Parabolic Mirrors

Take the model from `parabola_slider.scad` again and consider the line coming down to the curve from the top. If the parabola was a mirror and the string was a ray of light, the light would reflect at the same angle at which it hit the parabola (as we saw back in Chapter 8 when we talked about reflection). This means that the ray of light would next travel to the focus. In the case of our model, the light ray would follow the string. If you were to measure the angle the string coming into and going out of the slider makes with the parabola curve, you would find that they are equal (Figure 10-22).

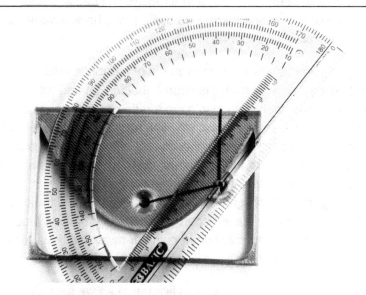

FIGURE 10-22

Measuring from the slider to the directrix

Similarly, if you had a bright light or a source of sound coming from the focus, it would put out a nice parallel set of light or sound waves. You can see this with the model by examining how an incoming ray perpendicular to the top will always bounce to the focus. This is illustrated by Figure 10-23.

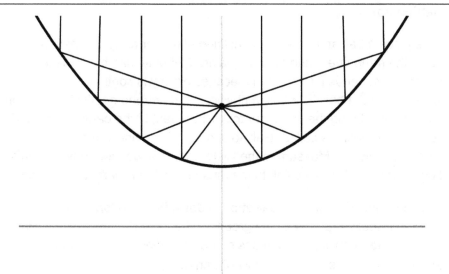

FIGURE 10-23

Light rays into or out of a parabola

Parabolic reflectors to focus light or sound are found in many places. For example, satellite TV dishes often have parabolic cross-sections. Things get a little more complex in 3D, and a receiver placed at the focus will block some of the reflector, so these dishes may have a more complex shape to optimize around that.

Flashlight reflectors are another example of parabolic reflectors. Some light will be cast directly from the source (bulb/LED) and will disperse quickly, but many flashlights also have a reflector that, to varying degrees, approximates a parabola. This is why you often see a brighter but narrower inner beam. The more closely the reflector approximates a parabola (and the closer the light source is to its focus), the narrower the beam will be, and the less fall-off in intensity you will get with distance. Special long-throw flashlights are designed with highly parabolic reflectors and carefully-placed sources, so that light rays will be very parallel, and you can brightly light a point far away, but those beams will never illuminate a very wide area.

There are whispering gallery effects like the ones we saw with ellipses from strategically-placed parabolic walls, too. Sometimes called *acoustic mirrors*, curved walls or dishes can be used to collect and focus sound. Giant concrete acoustic mirrors were put in place along the south coast of Britain in World War II to detect the sound of enemy bombers before radar came along and made them obsolete.

Galileo Ramp

If we place a ball at the top of an inclined plane, or ramp, the ball will roll down the ramp due to the force of gravity. Not only that, but it will also go faster and faster as it rolls down, because gravity is continuously accelerating the ball. The Italian physicist and astronomer Galileo is often credited with being the first to try this experiment in the early 1600s, and so this is sometimes called "Galileo's ramp." If you find yourself in Florence, Italy, the Museum of the History of Science there has Galileo's original ramp. We have a link to the museum in the References section.

Now, if we just dropped the ball straight down from the top of the ramp, how long would it take to hit the ground? The force due to gravity pulling down on the ramp is just the mass of the ball times the acceleration due to gravity, g. This is $9.8 m/s^2$ (on Earth, anyway).

If it was simply falling from h, the time to reach the ground (for reasons you can see in our *Make: Calculus* book) is

$$h = \frac{1}{2}gt^2, \text{ so}$$

$$t = \sqrt{\frac{2h}{g}}$$

What that means is that for every second the ball is in flight, gravity drags it down at a speed that increases by 9.81 m/s *every second*. Note that $t = 0$ is the time at which we drop the ball. If the height was 10 m, you would find out that the time for it to fall (in seconds) is

$$t = \sqrt{\frac{2*10}{9.81}} = 1.43$$

But now we are not dropping the ball. Instead, the ramp will support the ball as it falls, and it will also be traveling horizontally. For a ramp that is inclined at an angle θ to the ground, in effect we want to think about the acceleration parallel to the surface of the incline (Figure 10-24), which is $\sin(\theta)\,g$. So, the distance along the plane as a function of time will be

$$x = \frac{1}{2} \sin(\theta)gt^2$$

Solving for time t we get

$$t = \sqrt{\frac{2x}{g \sin(\theta)}}$$

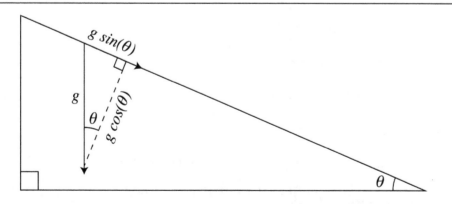

FIGURE 10-24

The Galileo ramp

If you search on "Galileo ramp" or "inclined plane experiment" you will discover a variety of ways of implementing a ramp to minimize friction. Then you can compare theory to experiment by measuring how long it takes to get to different points on the ramp versus how long our last equation says it should take. We particularly liked this Instructable: **instructables.com/Galileo-Ramps-exploring-velocity-and-acceleration-/** We have tried this with kids stationed along a ramp with stopwatches to measure how long it takes to get to a point on the ramp. It is challenging to get reasonable measurements that way, though, even if you set up a ramp that is 10 or 20 feet long, since the marble or toy car goes by fast.

Finding a Maximum or Minimum

One handy thing about a parabola is that it is easy to find out where it has a maximum or a minimum. We know that the vertex will be at $\frac{-b}{2a}$, and we can plug that back into the equation for the parabola to get the y value of the maximum or minimum.

Suppose we have developed a communication system, and we are going to charge customers to transmit messages that they send to one another. In this simplified scenario, there would be some initial cost for us to build the network, and users would pay us monthly subscription fees. However, more users on the network means there will be more traffic, since each user has more people to talk to. In fact, the traffic would be roughly proportional to the square of the number of users, since each user would generate an amount of traffic proportional to the number of other users they can talk to.

So let's say:

initial cost = 10,000

x = number of subscribers

income from subscribers = 50x

cost of managing subscriber traffic = $0.01x^2$

Putting that all together, if profit equals income minus costs

$$\text{profit} = 50x - 0.01x^2 - 10,000$$

The vertex is at $\frac{-b}{2a} = \frac{-50}{2 * -0.01} = 2500$ and our profit at that point is

$$50 * 2500 - 0.01 * 2500^2 - 10,000 = 52.500$$

So we would want to keep our network small (2,500 users) or get our costs down per user, or limit how many other users each one can really interact with.

This is called a *minimax problem* by some authors. Although this is a simplistic problem with made-up numbers, you can imagine that in real life it can be quite useful to find optimum values for systems that have quadratic terms.

Terminology and Symbols

Here are terms we used in this chapter which you might want to look up for more information:

- ± (plus and minus)
- completing the square
- coefficients
- directrix
- factoring
- focus
- Fundamental Theorem of Algebra (also d'Alembert's Theorem)
- minimax problem
- parabola
- polynomial
- quadratic equation
- quadratic formula
- roots of an equation
- spurious (or extraneous) solution
- vertex (plural vertices)

Chapter Key Points

In this chapter, we learned about the properties of a parabola and how to graph one. A given point on a parabola is an equal distance from the focus and the directrix. We learned how to graph a parabola, and the equation that governs a parabola. Through the technique of completing the square, we discovered the quadratic formula, which allows us to solve quadratic equations. We developed some intuition about how the graphs of parabolas change as the coefficients a, b, and c are varied. Finally, we saw where parabolas appear in real life, and the applications of being able to analyze them. In the next chapter, we cover the last of the sections, the hyperbola.

References

Here are a few sources for you to explore this topic in more depth:

The Wikipedia article "Parabola" is very good as a starting point. Derivations in the **Khan Academy** (http://khanacademy.org) go at the material in this chapter a little differently than we did if you would like alternatives.

The Galileo Science History Museum in Florence, Italy, has Galileo's original instruments in its collection, and photos of some of them on **its website** (http://museumsinflorence.com/musei/ History_of_Science_museum.htm).

Merzbach, U. C., and Boyer, C. B., (2011) *A History of Mathematics* (3rd ed.). John Wiley and Sons. This book is a broad resource on the worldwide history of mathematics. It is written assuming that the reader knows mathematical terminology at the college level but is a very comprehensive guide to use as a reference.

11 Hyperbolas

3D Printable models used in this chapter

See Chapter 2 for directions on where and how to download these models.

- `hyperbola_slider.scad`
  - This model creates a hyperbola given the distance between its vertices and foci, and other parameters
- `conic_section.scad`
  - This model creates a double-cone hyperbola given the distance between its vertices and foci, and other parameters
- You will also need
  - Elastic cord. We used 0.8 mm elastic cord, also called beading thread or crafting cord. You will need about 1 foot per model you choose to make.
  - A ruler or calipers, and a protractor
  - A few rubber bands

The final conic section is a result of cutting through a cone at an angle greater than the slant angle. This cross-section is a *hyperbola*. A hyperbola is a pair of curves (often called *branches*) that do not touch each other but go off to infinity in opposite directions. To see the entirety of a hyperbola you need two cones, placed vertex to vertex.

Hyperbola Conic Section

The simplest way to create a hyperbola is to cut a cone at 90° to its base. Any such cut that does not go through the vertex will result in a hyperbola. Let's use the model `conic_section.scad` (originally introduced in our book *Make: Geometry*) to see what this looks like. Figure 11-1 illustrates the geometric parameters that we need to set in conic_section.scad . (Figure 11-1 shows the cut for an ellipse, but the parameters are used the same way here.) They are

- r , the radius of the cone, in mm
- h , the height of the cone, in mm
- slicetilt , the angle of the one about which the cutting plane is rotated, in degrees
- Two parameters for the position of the axis about which the cutting plane is rotated:
  - sliceheight , in the z direction (mm), and
  - slicehoffset , in the plane of the base of the cone (mm).

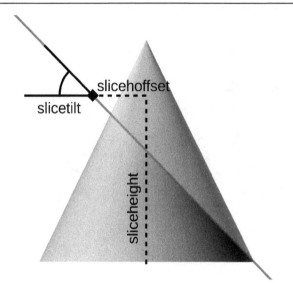

FIGURE 11-1

Definitions of the parameters of `conic_section.scad`

Table 11-1 lists the parameters used to create the pair of cones in Figure 11-2. We have to run the model twice, once to generate each cone. In this case, though, the parameters are the same for both cones. This produces a cut halfway between the center of the cone and its edge. As you can see in Figure 11-2, the cuts are symmetrical about the axes.

Table 11-1. Parameters for a hyperbola cut at 90° (Figure 11-1)

Model parameter	Both left and right cone
h	70
r	h / 2
slicetilt	90
sliceheight	0
slicehoffset	r / 2

FIGURE 11-2

Hyperbola cross-section across the pair of cones cut at 90°

It gets a little trickier if we want to try out making a hyperbola with a plane that cuts at an angle other than 90°. Then, the cutting plane will intersect the bottom and top cone at the same angle but create a different cross-section in each cone and exit along a different line through their respective bases.

In the example we will go through here (and that is 3D printed with red translucent filament in Figure 11-3), the angle of the cutting plane is 80°. For that example, once again the height (h) is 70 mm, and the radius (r) is h / 2 .

On the lower cone, we used an offset of $\frac{r}{8}$ to the left of the centerline for the axis of rotation on the cutting plane but kept it at the height of the bottom of the cone, or `sliceoffset = 0`. The centerline of the cone is an offset of 0, so this worked out to make the variable `slicehoffset = -r / 8`.

Thus, we have our plane of rotation rotating about a line in the plane of the base of the lower cone, and `r / 8` offset to the left of center. If we want to define the same plane from the point of view of the top cone (in other words, drawing the cutting plane and looking at the pair upside down) this same cutting plane will look like a plane at height `2 * h`, and offset `r / 8` to the *right* (and thus of the same value, but the opposite sign) as the offset of the lower cone.

The red cones in Figure 11-3 illustrate this example. The hyperbola intersections are a little different from each other. The branches of a hyperbola will be symmetrical in an infinitely long pair of cones, but since the cones here are finite, we see slices that are a little different from each other in the two cones. The parameters used for the two cones are summarized in Table 11-2. Once again, we need to run `conic_section.scad` twice to create the two cones.

FIGURE 11-3

*Hyperbola with 80°
cut*

Table 11-2. Parameters for a hyperbola at 80°

Model parameter	Left cone	Right cone
h	70	70
r	`h / 2`	`h / 2`
`slicetilt`	80	80
`sliceheight`	0	`2 * h`
`slicehoffset`	`-r / 8`	`r / 8`

If you want to try creating a different version of these cones, you may find it helpful to draw two copies of the diagram in Figure 11-1 to work out the values. Also, remember that `slicetilt` has to be greater than the slant angle of the cone to produce a hyperbola. It is very easy to mess up the signs of the offsets, so you may want to figure them out by walking through the geometry in detail as we did. For many values of `slicetilt` for a given cone, there will be no intersection between the plane and one of the cones, since the intersection would take place farther down the (hypothetical) infinitely long cone. Work out the details of both cones before printing one of them.

Graphing a Hyperbola

Now that we know what one looks like as a conic section, we need to put on our analytic geometry hat and try to figure out what the equation of this thing might be. We can think of a hyperbola as sort of an ellipse flipped inside-out. Like an ellipse, a hyperbola has two foci. A hyperbola has two *branches,* the open ends of which head out to infinity. Like a parabola, each branch of a hyperbola has a *vertex,* a point where each of the two branches comes closest to the other.

As the hyperbolas go farther and farther from the vertices, they approach a pair of straight lines, called the *asymptotes.* The asymptotes cross at the point midway between the vertices of the two branches. Figure 11-4 shows axes that cross at the midpoint between two branches of a hyperbola. The foci are the dots. The diagonal lines crossing through the midpoint are the asymptotes. You can see that each branch of the hyperbola is getting closer to the asymptotes as it goes away from its vertex.

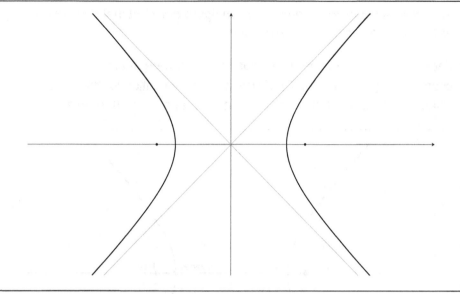

FIGURE 11-4

Graph of a hyperbola

Foci and Directrix

As we move along the curve of a particular hyperbola, the distance to each of the two foci will change. However, the *difference between* those two distances remains constant. As we learned in Chapter 9, the distances to the two foci of an ellipse sum to a constant, so this is sort of an inside-out ellipse. We will call the distance from the farther focus to our point on the hyperbola d1, and the distance from the nearer one, d2:

$$d1 - d2 = \text{constant}$$

We can rearrange this a little to say instead

$$d2 = d1 - \text{constant}$$

How can we subtract a constant offset from the distance to the farther-away focus? We need to apply this offset no matter what direction we are measuring our distance from. If we can, we could do something similar to what we did with a parabola in Chapter 10.

One way to add an offset to the distance to a point is to draw a circle around that point, and measure d1 minus the radius of that circle. This circle is called a *circular directrix*, in contrast to the straight-line directrix we encountered in Chapter 10. We can create a model very similar to the one we made for the parabola, but now instead of being a line, the directrix is a circle centered on one of the foci. Since the circle must be the same radius everywhere, if we can figure out what that constant is anywhere, we can know it everywhere.

Let's say we have a hyperbola whose vertices were at the x, y coordinates $(a, 0)$ and $(-a, 0)$. Following our convention for the ellipse in Chapter 9, let's say that our foci are at $(c, 0)$ and $(-c, 0)$ (Figure 11-5).

FIGURE 11-5

Definitions of dimensions of a hyperbola

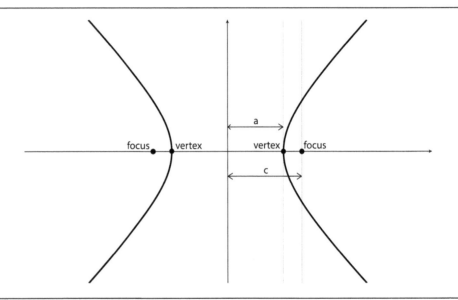

If the distance to one focus, minus the distance to the other focus, is a constant, let's see what that constant is at a vertex.

- The distance from the vertex at $(a, 0)$ to the focus at $(c, 0)$ is $c - a$.
- The distance from the vertex at $(a, 0)$ to the focus at $(-c, 0)$ is $c + a$.
- Therefore, we know that $(c + a) - (c - a) = \text{constant} = 2a$.

And so, since this is true in general, the radius of our circular directrix is $2a$, or exactly the distance between the two vertices. (Remember that the circular directrix is, however, centered at a focus, not at a vertex.)

The Hyperbola Model

The model `hyperbola_slider.scad` allows you to slide along the hyperbola and see how these two distances stay the same as each other. Unlike the ellipse, the sum of these distances is not a constant — the *difference between* them is a constant. As we will see in the section about the equation of a hyperbola, the standard way to write one, like an ellipse but with a negative sign in front of the y^2 term, is

$$\frac{x^2}{a^2} - \frac{y^2}{b^2} = 1$$

where $b^2 = c^2 - a^2$, the difference of the squares of twice the spacing between the vertices and twice the spacing between foci. (Where this all comes from will make more sense once we play with the model in subsequent sections.) The model has the following parameters you can set; the defaults are shown here.

- `a = 10;`
  - location of vertices (separation is $2a$) in mm
- `b = 25;`
  - Hyperbola scaling parameter as defined in previous paragraph, mm
- `wall = 2;`
  - width of the directrix and its support, mm
- `holesize = 5;`
  - size of holes for focus and slider, mm
- `h = 8;`
  - overall thickness of the model, mm
- `asymptotes = 1;`
  - notch size for asymptote string, mm (see later section on asymptotes)
- `size = 100;`
  - Radius of outer circle, mm

Do Not Shrink This Model

This model will be easier to handle if it is as large as your printer will permit. We recommend against going much smaller than the default dimensions, because the slider may get too small to work well.

Assembling the Model

Like the ellipse model in Chapter 9 and the parabola in Chapter 10, some assembly of this model is required. 3D print the model and be sure you do not lose the small part (the slider). Cut a piece of stretchy cord a little more than twice the maximum radius of the model. Take the cord and run it through the hole at the parabola focus from back to front (Figure 11-6). Then run the cord through the slider from front to back (Figure 11-7). Next take the cord out along the back of the model to the outer radius (rounded long edge) as we see in Figure 11-8. Finally, as shown in Figure 11-9 pop the string over that edge of the model, and back up the front and through the slider again from front to back. Tie off the string on the back and cut off the excess.

FIGURE 11-6

Starting to connect slider and its track

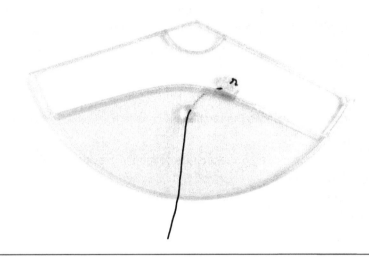

FIGURE 11-7

Looping around the back

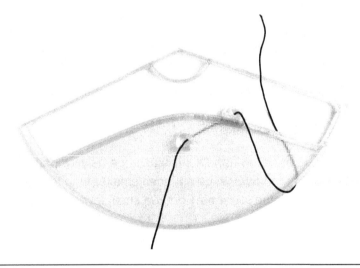

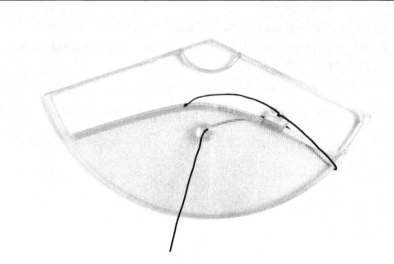

FIGURE 11-8

*Bringing the string
front again*

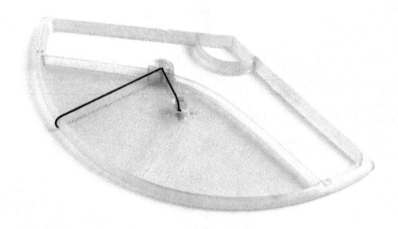

FIGURE 11-9

Finished model

You are done! For now, ignore the notches on the long straight bars that hold the directrix circle. Those are for the asymptotes. We will create those with rubber bands later.

Learning with the Model

If the model is assembled correctly, the string from the slider to the rounded edge on the right will lie along a radius of the circle centered at the center of the directrix (that is, centered at the other focus, on the left in Figure 11-10). If there is too much friction, the slider may be slightly out of position, but a little nudge should make it snap back into place.

Now you can measure the distance from the focus to the curve with a ruler (along the string, as shown in Figure 11-10) and the distance from the curve straight down to the directrix (use the vertical string as a guide, as in Figure 11-11). You will find they are equal.

FIGURE 11-10

Measuring distance from curve to focus

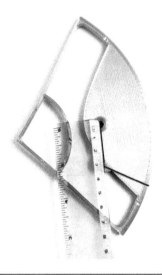

FIGURE 11-11

Measuring distance from curve to directrix

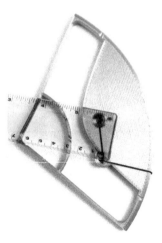

Similar to the parabola model in Chapter 10, the distance from the curved edge of the model to the circular directrix stays constant. In this case the curved edge and circular directrix are concentric circles. Therefore, given that the length from the focus to the slider stays the same as the distance from the slider to the directrix, we can deduce that the length of the string stays constant.

Now, move the slider around the hyperbola a bit farther, and convince yourself that these relationships hold there, too. Be careful that the string lies along a radius of the circle centered on the other focus.

Note that the model only shows one branch of the hyperbola. Both foci are visible: one is at the center of the circular directrix, the other at the hole for the string. In Figure 11-12, we show two copies of the model, lining up the foci and centers of the circular directrix of the models.

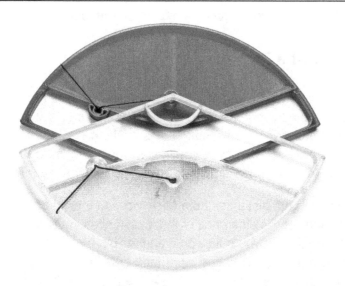

FIGURE 11-12

How models of two branches align

Model Limitations

What are the sources of inaccuracy in this model? To be accurate, the two parts of the string should emerge from the slider in the same place, which should be exactly on the curve. When the angle between them becomes large, this does not always work perfectly.

Also, when measuring the distance from the focus, you need to be careful to measure from the center of the hole, rather than measuring the length of the string on the surface. Making this hole smaller would make it easier to measure, but harder to assemble.

When the string stretches, its total length gets longer, which allows the angle to get slightly off. The parabola model had this problem too. String that does not stretch would be ideal, but we found that it was difficult to tie the loop tightly enough, and that the friction was too high when the string did not have any give.

When the value of c is much larger than b (so that the focus is far from the curve), the angles of the string stay relatively acute, which makes the model behave more accurately, but this also means that it is not possible to see as much of the hyperbola curve.

This model would be a little tricky to create without a 3D printer. However, you can do something philosophically similar with a ruler, some tape, and a string. Search on "constructing a hyperbola with string" for videos.

Equation of a Hyperbola

To continue with our study of analytic geometry, we need to figure out how to express a hyperbola as an equation, based on the definitions of its geometrical properties we have already seen in the chapter to this point. Deriving the equation of a hyperbola generates some messy algebra. Let's see if we can find as simple a way to show this as possible.

We need to find the distances from each focus to a point on the hyperbola and set the difference of these distances equal to 2a to get our equation. Imagine a hyperbola with branches opening to the left and right, and there is a point (x, y) on the right-hand branch of the hyperbola. We will assume this hyperbola is centered at $x = 0$ and $y = 0$, and the foci are at $(0, c)$ and $(0, -c)$, as we see in Figure 11-13.

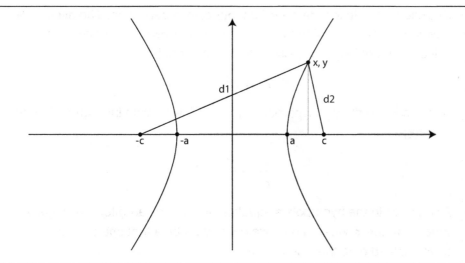

FIGURE 11-13

Distances to key points on a hyperbola

Then the distance to the farther away focus is

$$d1 = \sqrt{(x+c)^2 + y^2}$$

And the distance to the nearer focus is

$$d2 = \sqrt{(x-c)^2 + y^2}$$

By the definition of a hyperbola, and using our special case to get the value of the constant, we know that

$$d1 - d2 = 2a$$

And so, the equation of a hyperbola is

$$\sqrt{(x+c)^2 + y^2} - \sqrt{(x-c)^2 + y^2} = 2a$$

Now, some nasty algebra ensues, repeatedly rearranging and squaring both sides. We will not repeat that here, since it is spelled out well in other places. We particularly recommend the very careful narration of this derivation in the **Khan Academy** (https://www.khanacademy.org) video, "Proof of the Hyperbola Foci Formula." Sal Khan uses f where we use c, however. Since c is more common in books, we will stick with c here. The derivation is in many precalculus books and websites as well. When you get through getting rid of all the square roots, you are left with

$$a^2 - c^2 + (c^2 - a^2)\frac{x^2}{a^2} = y^2$$

This might start to look familiar from our ellipse derivation. The difference between the squares of the focus distance and vertex distance is usually called b^2, in analogy to an ellipse's semimajor axis:

$$b^2 = c^2 - a^2$$

So, we can rewrite our hyperbola equation, if we divide through by b^2 and rearrange, as

$$\frac{x^2}{a^2} - \frac{y^2}{b^2} = 1$$

Going back to the hyperbola's equation, if the y squared term is negative rather than the x squared one, we get a hyperbola that opens up and down, instead of to the sides. That is,

$$\frac{y^2}{b^2} - \frac{x^2}{a^2} = 1$$

will open up and down rather than right and left.

Asymptotes

It turns out that for both right-and-left and up-and-down opening hyperbolas, the asymptotes are the lines

$$y = \frac{b}{a}x$$

And

$$y = -\frac{b}{a}x$$

Adjusting the difference between the position of the focus and the vertex will change the slopes of the asymptotes and correspondingly squish or spread out the hyperbola. If the foci are much farther out on the axes than the vertices, then b will be large compared to a, and the asymptotes will climb steeply. If, however, the opposite is true, the asymptotes will be at shallow angles, and the hyperbola will have a narrower opening.

The model has notches to run string or two rubber bands to mark the asymptotes. The two asymptotes will cross in the open space between the vertex of the hyperbola corner that marks this branch of the hyperbola's distant focus. If you use the default values of a and b, there should be plenty of room to do that. However, you will need to park the slider at the vertex of the hyperbola to keep it from distorting the asymptote (Figure 11-14).

FIGURE 11-14

Placing rubber-band asymptotes on the model

Another Example

If we instead create a model with

```
a = 20;
b = 25;
```

this means that the focus and vertex are farther apart than for the example we have used so far, in which `a = 10` and `b = 25`. This means that the hyperbola is a little steeper. You can print one of these, too, to see how it works out. In this case, the asymptotes cross the directrix. There are notches in the directrix to allow the asymptotes to pass over correctly (Figure 11-15).

FIGURE 11-15

Hyperbola with larger value of the a coefficient than previous example

To test your intuition on the material so far, draw x-y axes on a piece of paper. Take two rulers and use them to represent the asymptotes, and sketch in the hyperbola for these equations:

$$x^2 - y^2 = 1$$

$$\frac{x^2}{9} - \frac{y^2}{16} = 1$$

$$\frac{y^2}{16} - \frac{x^2}{9} = 1$$

How are the asymptotes for the second and third equations related? The answers are illustrated in Figure 11-16. (They are the same.) The hyperbola corresponding to the second equation is shown on Figure 11-16 in red (opening left and right); the third, in the blue. **Asymptotes are black.**

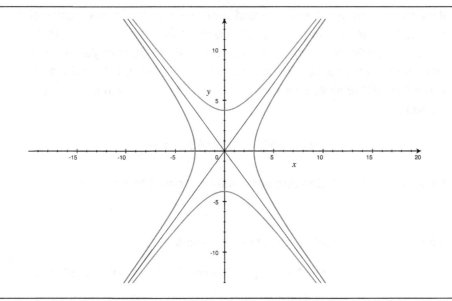

FIGURE 11-16

*Hyperbola and
asymptote examples*

Reciprocal Form

To this point, we have talked about hyperbolas that have squared terms in both x and y. However, there is an alternative (and somewhat surprising) form that also has all the properties of a hyperbola. It is

$$xy = c$$

or

$$y = \frac{c}{x}$$

where c is a constant. This is sometimes called the reciprocal function, since x and y are reciprocals of each other (possibly with a constant multiplier). It turns out that if we take a hyperbola where the a and b constants are equal, like

$$x^2 - y^2 = 1$$

and rotate it 45°, we get this formulation. In Chapter 8, we saw that rotating a coordinate system by an angle A uses this pair of formulas.

$$X = x \cos (A) + y \sin (A)$$
$$Y = y \cos (A) - x \sin (A)$$

If we want to know the sin(45°) and cos(45°), you could use a calculator, but using a decimal might mask the point we are about to make. Instead, since this is after all a trig book, consider a 45-45-90 triangle. The two sides need to be equal to each other (let's say of length 1) and that would make the hypotenuse, by the Pythagorean Theorem, equal to $\sqrt{2}$. That means that

$$\sin\,(45°) = \cos\,(45°) = \frac{1}{\sqrt{2}}$$

Let's try out this rotation and see what happens. Starting with

$$x^2 - y^2 = 1$$

substitute in the equation for rotated coordinates

$$[x \cos\,(45°) + y \sin\,(45°)]^2 - [y \cos\,(45°) - x \cos\,(45°)]^2 = 1$$

observe that there is a $\left(\frac{1}{\sqrt{2}}\right)^2$ factor on all the terms (which just equals $\frac{1}{2}$). We pull out that factor of $\frac{1}{2}$, and multiply out the squared terms. We get

$$\frac{1}{2}\left(x^2 + 2xy + y^2\right) - \frac{1}{2}\left(x^2 - 2xy + y^2\right) = 1$$

The squared terms in x and y fall out, and the $2xy$ terms add, so we get

$$\frac{1}{2}4xy = 1$$

or

$$2xy = 1$$

and

$$xy = \frac{1}{2}$$

As we can see in Figure 11-17, this is a hyperbola whose asymptotes are the axes. Note that this rotation only produces this tidy result if $a = b$. (In this case $a = b = 1$.) Otherwise, we wind up with a messy algebraic thicket, and the asymptotes are no longer the axes.

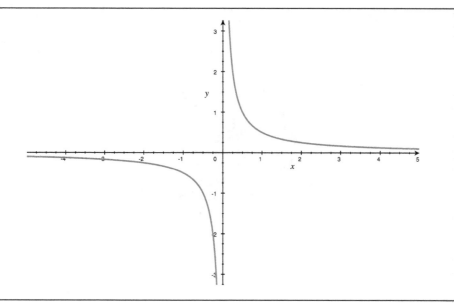

FIGURE 11-17

Graph of rotated hyperbola $xy = \frac{1}{2}$.

Hyperbolic Mirrors

Hyperbolas come up in optics quite often, merged with other conics to get rid of effects like the rainbow halos called *chromatic aberration* that otherwise appear around objects in a telescope. As we saw in Chapter 8, refraction of light through a lens is of necessity going to bend different colors of light a little differently. This makes it challenging to make very large telescope lenses that do not have this problem.

One way around it is to use mirrors in a telescope instead. A mirrored surface (for example, a parabolic one) can focus light just as a lens can, but with fewer problems since the light is not going through a thick lens. Most large modern telescopes are some sort of *reflector*.

When a light ray hits a mirror, it bounces off at an angle equal to the angle at which it hits the surface, as we learned in Chapter 8. We show an incoming and reflected ray in Figure 11-18. Explore the model and prove to yourself that any ray coming in from the right along a radius of a circle centered on the left focus will reflect to the right focus. If you have a protractor, try to measure the angles. Light emitted at the right focus will spread out on lines centered at the left focus.

FIGURE 11-18

Light reflecting from a hyperbola reflecting at same angle as incident light

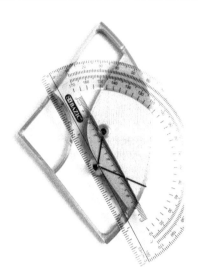

This assumes we are reflecting light from the concave surface, though. More commonly, hyperbolic mirrors are convex, and are secondary mirrors placed at the focus of a bigger, primary mirror. One arrangement of this is called a *Cassegrain* telescope, which is notable for being a very compact arrangement of optics. This allows a big telescope to fit in a modest dome (or be launched into space in a limited-volume rocket capsule). Remarkably, although very modern telescopes use the design, Laurent Cassegrain was active in the mid-1600s. He was the equivalent of a high school teacher in France.

The primary mirror of a Cassegrain is a paraboloid shape. Light rays enter the telescope from the sky as parallel beams, and, as we saw in Chapter 10, are collected at the paraboloid's focus. Before those light rays reach this focus, though, they hit the secondary mirror. The secondary mirror is a convex hyperbolic shape that shares a focus with the primary mirror (the blue dot in Figure 11-19).

This bounces light directed toward that focus to converge on the hyperbola's other focus (the red dot), through a hole in the primary mirror. Instruments, detectors, or (for small telescopes) eyepieces are usually placed at this *Cassegrain focus* (red dot in Figure 11-19). The position of the foci of the mirrors is tailored so that the secondary mirror will form images a bit behind the primary mirror.

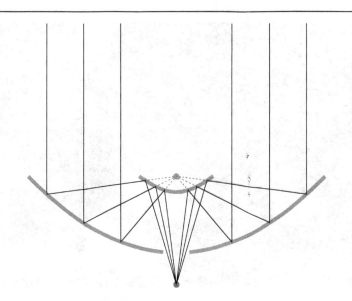

FIGURE 11-19

Cassegrain telescope light path

Actually designing and optimizing a Cassegrain telescope requires some sophisticated math, reaching up into calculus. If you want to appreciate a mostly-algebra discussion of how light reflects off a hyperbolic mirror, there is an explanation and interactive graphics on mathematician Carol JBV Burns' site, *One Mathematical Cat*. It is Lesson 157, "Reflecting Property of a Hyperbola," which we reference in full at the end of this chapter.

Many famous large telescopes are Cassegrains or variations on it, including Caltech's 200-inch diameter Hale telescope at Mt. Palomar in California. The James Webb Space Telescope (JWST) is fundamentally a Cassegrain too, although its primary mirror might not look like it. The primary mirror was created in honeycomb-like segments so it could be folded up for launch. The Cassegrain arrangement works for radio waves too. Large antennas of NASA's Deep Space Network, like the one in Figure 11-20 are also paraboloid primary dishes, with a hyperboloid secondary reflector. If both the primary and secondary mirrors are hyperboloids, it is called a *Ritchey-Chrétien* design.

FIGURE 11-20

A deep space network antenna. Courtesy NASA/JPL– Caltech

Hyperbolic Trigonometry

Back in Chapter 5, we learned about the unit circle. We found out that we could relate our sine and cosine functions of single triangles to a continuously varying angle and thus generate a circle. It turns out that there are other functions that let us do something analogous for a hyperbola. However, to show this, we need to tie together a few pieces from previous chapters and introduce some new ideas as well.

In Chapter 5, we learned that for a circle of radius equal to 1, we could express our x and y coordinates in term of an angle θ as follows:

$$x = \cos(\theta)$$

$$y = \sin(\theta)$$

$$\tan(\theta) = \frac{\sin(\theta)}{\cos(\theta)}$$

If we varied θ, then the resulting (x, y) values would run along a circle of radius 1. As we saw in Chapter 9, the equation of a circle is

$$x^2 + y^2 = r^2$$

If we insert $\cos(\theta)$ for x, $\sin(\theta)$ for y, and let $r = 1$ in our equation for a circle, we get the trig identity for squared sine and cosine, which we learned about in Chapter 6:

$$\sin^2(\theta) + \cos^2(\theta) = 1$$

For this reason, sine and cosine are sometimes called "circular functions," since they are related so closely to circles (as well as triangles, of course). So that might all lead you to wonder whether there might be other useful "unit conic sections." As it turns out, the unit hyperbola,

$$x^2 - y^2 = 1$$

has a different kind of trig functions associated with it. Not surprisingly, these are called *hyperbolic trig functions*, and are written sinh(x), cosh(x), and tanh(x). These are pronounced either as "hyperbolic sine (or cosine, or tangent) of x" or, more commonly, "cinch of x", "cosh of x" and "tanch (or "than") of x." Some parts of the world pronounce sinh as "shine" for some reason. Note the distinction between sinc(x), usually pronounced "sink," which is $\frac{\sin(x)}{x}$, versus sinh(x).

For this analogy between circles and hyperbolas to hold we need to express x and y in terms of our hyperbolic functions somehow. We would move along our hyperbola as we vary some third variable, say, p, equivalent to varying θ in the case of sine and cosine. As it turns out, the hyperbolic functions meet this criterion if we let

$$x = \cosh(p)$$

and

$$y = \sinh(p)$$

Not surprisingly, just as $\tan(x) = \frac{\sin(x)}{\cos(x)}$, $\tanh(x) = \frac{\sinh(x)}{\cosh(x)}$.

Okay, but what are these functions, and how do we compute them? It turns out that these functions are not tied to triangles. They are instead tied to *exponentials*, functions that rise (or fall) very rapidly. Back in Chapter 6, we learned about logarithms, and the inverse of taking a log of raising 10^x. The function $y = x^2$ is an exponential function.

But one could use any number as a base. As it turns out, there is a different number that comes up all the time in physical applications called *Euler's number*, usually written with a lowercase "e". (Named after the same Euler we have met in other chapters.) It is equal to approximately 2.71828. Like pi, it is a fundamental constant that comes up all over the place. As pi is to a circle, so *e* is to hyperbolas and things that grow and shrink rapidly. We spend a lot of time in the company of this number in *Make: Calculus*.

Logarithms to the base e are usually referred to as "natural logs", with the function being written ln(x). If we talk about an "exponential function," typically we are talking about one with e as the base, not 10. Just as $\log(10) = 1$, $\ln(e) = 1$, and $e^1 = 2.71828$.

What does that have to do with our cosh and sinh? It turns out that our x and y variables that will draw out our unit hyperbola are functions of e. Analogous to having x and y be functions of the angle θ, here our x and y are functions of a third variable, p, to make the point that we vary p, and the resulting x and y values will lie along a unit hyperbola.

$$x = \cosh(p) = \frac{e^p + e^{-p}}{2}$$

$$y = \sinh(p) = \frac{e^p - e^{-p}}{2}$$

Note that as we increase our variable p, our values of x and y increase or decrease rapidly. Unlike circular trig functions, hyperbolic trig functions are not periodic. Also, x can never be negative since e raised to a power is always positive, so we are only able to sweep out half of our hyperbola with these functions. This is a right-and-left opening hyperbola, so we are only generating the right branch.

To put it another way, instead of going around a circle, our x and y variables sweep us off on part of our hyperbola. Figure 11-21 shows graphs of $y = \sinh(x)$, which is the red curve rising steeply from negative values; $y = \cosh(x)$, the blue curve that sort of looks like an upward-opening parabola; and $y = \tanh(x)$, the green curve that rises more slowly than $\sinh(x)$.

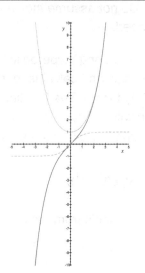

FIGURE 11-21

*Graph of sinh(x)
(red) cosh(x) (blue)
and tanh(x) (green)*

If you were to do the (rather messy) algebra of using these values of x and y in the equation

$$x^2 - y^2 = 1$$

we can replace x and y to get

$$\sinh^2(p) - \cosh^2(p) = 1$$

which in turn is

$$\left(\frac{e^p - e^{-p}}{2}\right)^2 + \left(\frac{e^p + e^{-p}}{2}\right)^2 = 1$$

you would find that, after a lot of terms canceled out, you would indeed wind up with

$$\frac{4}{4}e^p e^{-p} = 1e^{p-p} = 1e^0 = 1$$

since any number raised to the zero power equals 1. There are also inverse hyperbolic functions, $\sinh^{-1}(x)$, $\cosh^{-1}(x)$ etc. In parallel to the circular trig functions there are also hyperbolic secant, sech(x); cosecant, csch(x); and cotangent, coth(x). But let's not get too carried away.

The hyperbolic trig functions also have their own versions of trig identities, just like we saw for the circular trig functions in Chapter 6. Some of the relationships do work out to be a little different than their circular function analogs. Do not assume they are exactly the same; look up the relationships you need.

Finally, in case you were wondering, hyperbolic trig functions come up frequently in advanced physics and calculus, particularly for people studying things like relativity theory. The function e^x is very convenient in calculus problems, as you will find out if you continue to *Make: Calculus*, so this seemingly messy function is a real bonus in those contexts.

Terminology and Symbols

Here are some terms and symbols from the chapter you can look up for more in-depth information:

- asymptotes
- Cassegrain telescope
- directrix
- Euler's number (e)
- focus (foci)
- hyperbola
- hyperbolic trig functions ($\sinh(x)$, $\cosh(x)$, $\tanh(x)$)
- reciprocal function
- vertex (plural vertices)

Chapter Key Points

In this chapter, we learned about the final conic section, the hyperbola. It is the only conic section that crosses a pair of point-to-point cones, creating one branch on each of the cones. We saw that it can be modeled with respect to two foci and a circular directrix. We derived two forms of the equation for a hyperbola. We explored their properties and saw how hyperbolic mirrors enable creation of large telescopes. Finally, we learned about the hyperbolic trig functions, and the analogies to the ones derived from a circle.

References

For more background, the "Hyperbola" article on Wikipedia is a good starting point. To walk through all the steps of the algebra required to get the equation of a hyperbola, go to the **Khan Academy online** (https://khanacademy.org) and search on "Proof of the Hyperbola Foci Formula."

There are also more sophisticated effects of reflection from a hyperbolic surface in Nils Berglund's YouTube video, "**A hyperbolic wave reflector** (https://youtu.be/9VDU0FcgNLY)."

Finally, mathematician Carol JBV Burns' site, **One Mathematical Cat** (https://onemathematicalcat.org/Math/Precalculus_obj/tableOfContentsPreCalculus.htm), has many lessons that have alternative views of topics in this chapter and book, in many cases with interactive diagrams.

If you want to learn more about hyperbolic trigonometric functions, again the Khan Academy has good resources. Some YouTubers also have developed videos about various properties. Mathematician (and tech reviewer for this book) Niles Ritter has interesting insights about pi, e, circular and hyperbolic trig functions, and more in a **2016 blog post** (https://nilesritter.com/wp/?p=1816), "From Euclid to Euler to Einstein."

12 Applications and Looking Ahead

3D printed models used in this chapter

See Chapter 2 for directions on where and how to download these models.

- `simple_arm_sg90.scad`
    - OpenSCAD model for robot arm parts
- There are also a few very short 3D OpenSCAD models in the chapter you can type into OpenSCAD, and a link to a third-party repository for one of the models.

Arduino Sketch

- `servo_arm.ino`
    - Arduino sketch (not 3D printed model) to run on robot arm. It is also in the repository with the 3D printable models.

Other materials used in this chapter

- Robot project (as described in the chapter, but there are many ways to build this):
    - Three SG90 hobby servos with cross-shaped horns
    - One Arduino-compatible microcontroller board (e.g. Arduino Uno)
    - One PCA9685-based PWM servo driver board
    - A USB cable for programming, with the appropriate connector for your Arduino board
    - 4-AA battery holder with connections for screw terminals
    - Four rechargeable AA batteries (or three fresh alkaline ones)
    - Screws included with the servos
    - A set of breadboard jumper wires, male to female
    - Small cross-head (Phillips) screwdriver (size 00 to 1)
    - Small flat-head screwdriver (about 2.5 mm wide or less) for screw terminals
    - Possibly tools for cutting and stripping small wires (see narrative)

In this final chapter, we have two open-ended projects that use some of the mathematics we learned in this book. They are meant to be starting points that you can build upon, rather than step by step guides.

First, we have a tiling project, in which we figure out how to generate polygons that will completely cover a space, including a recently discovered one called an *einstein github*. Beyond giving you decorating ideas for the next time your family remodels a bathroom, playing with tilings can give you insights into the relationships among polygons and an opportunity to exercise single-triangle trig ideas. We include OpenSCAD code for creating 3D printable and 2D paper shapes.

Next, we explore how trigonometry can position a robot arm, using the mathematics of *inverse kinematics*. This is an intermediate electronics project, and to build it you will need to know how to use an Arduino microprocessor and servos. If you do not, you can read through the example and see the mathematical points of the exercise, even if you do not want to delve into the actual build. We give pointers to other sources if you do want to learn enough to give it a shot.

We include general references for the book as a whole at the end as well. Some have been noted in earlier chapters, but others may serve as broader backup to help you fill in any gaps. If you are planning to teach using this book, you might find some of the references useful to help think about teaching math hands-on in general. Following this chapter is a brief Appendix that lists the topics covered by chapter, in case you are a teacher trying to align this book to a traditional syllabus. But we are getting ahead of ourselves. Let's try some projects!

Tilings

The first thing that comes to mind with the word *tiling* (for non-mathematicians, anyway) is a bathroom floor, like the one in Figure 12-1. However, coming up with shapes that will cover a flat surface in an efficient and visually pleasing way is not as easy as it seems. In this section we will play with tilings, from simple polygons to a newly discovered way to cover a surface called an *einstein monotile* (no relation to the physicist).

FIGURE 12-1

Hexagonal tiles

Tiling, to a mathematician, involves covering a surface completely with one or more repeating shapes, such that the shapes do not overlap or have any gaps between them. It might seem that it should be possible to tile a flat surface with any regular polygon, but that is not the case. Only triangles, squares, and hexagons can tile without other shapes taking up the spaces in-between. Pentagons and shapes with seven or more sides cannot tile by themselves. To understand why, let's remind ourselves about the features of a regular polygon.

Tilings with Regular Polygons

Figure 12-2 shows a regular hexagon. We can construct the right triangle shown there from the hexagon's radius (a line from the center to a vertex), the apothem (a line from the hexagon's center which is perpendicular to a side), and half of one of the hexagon's sides. The hexagon can be broken up into 12 of these triangles: two for each of the sides.

Since all the triangles are identical, the interior angle of each will be $\frac{360°}{2*6}$, or 30°, as shown on the right of Figure 12-2. We know that the other angle of the triangle is a right angle, and the angles all need to add up to 180°. So that means that the angle that is half the vertex angle of the hexagon is $180° - 90° - 30° = 60°$. The total angle inside each vertex, then, is twice that, or 120°.

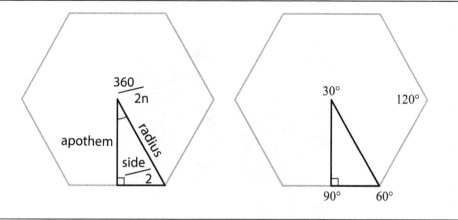

FIGURE 12-2

Laying out a hexagon

If we look back at our floor tiles in Figure 12-1, we see that each vertex is the intersection of three hexagonal tiles. We could draw a circle around each vertex and see that the angles have to add up to 360° (Figure 12-3). Thus, in the case of a hexagon, where each vertex angle is 120°, we can fit three of them inside the circle perfectly. Square tiles (Figure 12-4) have vertex angles of 90°, so four of them come together at each vertex. Six triangles (Figure 12-5) come together at each vertex in a triangular tiling.

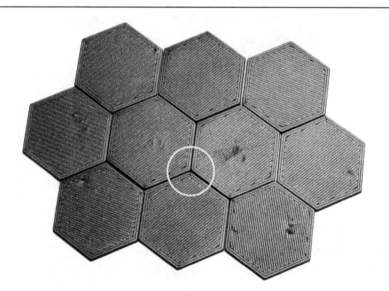

FIGURE 12-3

Hexagonal tiles with angles adding up at a vertex

FIGURE 12-4

Squares adding up at a vertex

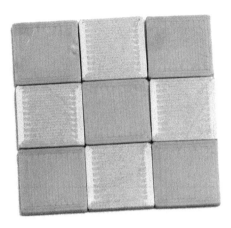

FIGURE 12-5

Triangular tiling

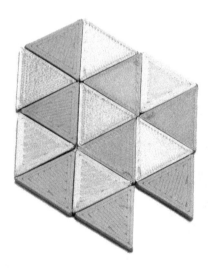

But what happens when we get to five sides (a pentagon)? The interior angle is $\frac{360°}{2*5}$, or 36°. That means that half the vertex angle is $180° - 90° - 36° = 54°$, and the vertex angle is twice that, or 108°. That means that 3.3333 pentagons would fit around a vertex, so it does not work. There is a good in-depth discussion of this in the 2017 *Quanta* magazine article by Natalie Wolchover referenced at the end of this chapter. Figure 12-6 shows one problematic tiling attempt.

FIGURE 12-6

Tiling attempts with pentagons

The formula for the vertex angle in general, for a regular polygon of n sides, is

$$\text{vertex\_angle} = 2\left(180° - 90° - \frac{360°}{2n}\right)$$

As we saw, a hexagon does happen to work out. For regular polygons with more than six sides, though, a mix of other shapes (which might not be regular shapes, with all their sides the same) are needed to create a tiling. Table 12-1 lists values of vertex angles for polygons with three to ten sides.

Table 12-1. Vertex angles of regular polygons

Polygon	Sides	Vertex angle (degrees)
Equilateral triangle	3	60
Square	4	90
Pentagon	5	108
Hexagon	6	120
Septagon	7	128.57
Octagon	8	135
Nonagon	9	140
Decagon	10	144

Now, if we no longer limit ourselves to tilings with all the same shape, and allow more than one, more possibilities open up. For example, suppose we used one square (90°) plus two octagons ($135° * 2 = 270°$). That adds up to 360°, and we can see in Figure 12-7 that a pattern of two octagons and one square (all with the same length sides) will meet at the vertices exactly. This is a common commercial tiling pattern too — watch for it now that you know about it! Of course, we can also come up with triangles or other shapes that no longer have all the sides of the same length to try and make tilings work.

FIGURE 12-7

Octagon with squares

Creating sets of these shapes to play with is easy with OpenSCAD. To print just one regular polygon of thickness t, with n sides of length s millimeters, just type this into OpenSCAD. (Set `t = 0` if you want to export an .svg file to laser cut or print in a 2D printer.)

```
s = 20; // length of a side, mm
n = 3; // number of sides
t = 2; // thickness in mm - set to 0 for 2D
if(t) cylinder(r = s / (2 * sin(180 / n)), h = t, $fn = n);
else circle(s / (2 * sin(180 / n)), $fn = n);
```

If you want to make multiple copies of one polygon in one output file, use the following instead. The parameters x and y are the maximum dimensions (in mm) we want to cover with our polygons for printing. To print them out on paper, U.S. copy paper is 8.5 by 11 inches, which gives you a printable area of roughly 200 by 260 mm. To print on paper, as before set `t = 0`. For 3D printing, set x and y to the dimensions of the printable area of your print bed (or smaller). This model will lay out a reasonably efficient grid of shapes, with a bit of margin, into the specified space.

```
s = 20; // length of a side, mm
n = 4; // number of sides
t = 2; // thickness in mm - set to 0 for 2D
x = 200; // x dimension of paper or print bed
y = 200; // y dimension of paper or print bed
for(
  x = [0:s / sin(180 / n) + 1:x - s / sin(180 / n)],
  y = [0:s / sin(180 / n) + 1:y - s / sin(180 / n)]
)
  translate([x, y]) rotate(180 / n)
    if(t) cylinder(r = s / (2 * sin(180 / n)), h = t, $fn = n);
    else circle(s / (2 * sin(180 / n)), $fn = n);
```

Print yourself a set of various polygons and try out tilings. Table 12-1 might help with your brainstorming. For example, two pentagons and a decagon meeting at each vertex we might expect to work, based on the sum of the angles. However, if we try it (Figure 12-8) it turns out that we wind up with a ring of pentagons around each decagon, with gaps between the pentagons. It is necessary for the vertex angles of tiles coming together to create a full circle, but that alone does not guarantee that those shapes can tile a whole surface without gaps. Other things are possible by mixing triangles, squares, and hexagons (Figure 12-9).

FIGURE 12-8

*Pentagon and
decagon tiling*

FIGURE 12-9

*Triangle, square,
and hexagon tiling*

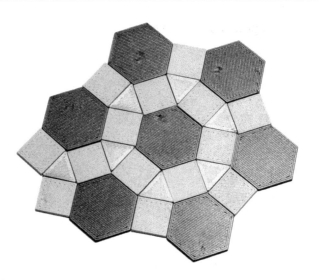

Similar repeated patterns arise in many other contexts, too. Close to home for us, hexagonal infill (Figure 12-10) is a strong and stable way to create the structure of a 3D print. This was a square part printed with zero top and bottom layers, and 20% hexagonal infill. And of course, a beehive is a closely packed set of hexagons.

FIGURE 12-10

Hexagonal infill

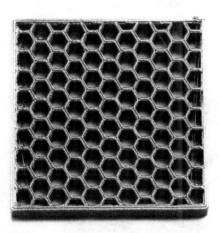

If you find this process of packing simple shapes into complex repeating patterns intriguing, you might also want to explore the different types of crystals, which pack molecules into 3D shapes. As we explored in our 2016 Apress book, *3D Printed Science Projects*, water ice crystals can be hexagonal in shape. Look up "crystal structures" to learn about these different types of repeating three-dimensional lattices.

Aperiodic Tilings

Not every tiling is a pattern that repeats. If a tiling is a result of a set of shapes that repeat over and over but do not form a repeating pattern, this is called an *aperiodic tiling*. There are a number of these, and it has been a bit of a mathematical competition for centuries to come up with interesting tilings that use very few pieces over and over. Many of these have both concave (cupped inward from the polygon's surface) in addition to the convex (bulging outward) shapes we are used to seeing in our regular polygons. For example, one type of Penrose tilings uses just two shapes, called the dart and the kite, to cover a surface aperiodically.

However, a holy grail for some time was to find an *einstein* tile. The name comes from the German phrase for "one stone" (not the famous physicist). An einstein, or *aperiodic monotile*, would tile a whole surface in a way that did not repeat. In 2023 the team of David Smith, Joseph Samuel Myers, Craig S. Kaplan, and Chaim Goodman-Strauss discovered an einstein that can cover a surface without ever repeating a pattern (albeit needing to be flipped over in some places). They dubbed it the "hat" (Figure 12-11) although some people think it looks more like a t-shirt.

FIGURE 12-11

One "hat"

You can read about the hat and the effort to prove that it does what the authors think it does on their **University of Waterloo site** (https://cs.uwaterloo.ca/~csk/hat/), which as of this writing has links to a preprint paper and other resources. English mathematician Christian Lawson-Perfect has created a repository of various ways to print out the hat and other tiles, in OpenSCAD or a variety of 2D and 3D printable formats. The models in the next figures are printed from his **Github repository's** (https://github.com/christianp/aperiodic-monotile) OpenSCAD file. There are other variants to explore discussed at the links, including one that does not have to be flipped over to tile the whole surface.

The hat has 13 sides, and is the irregular shape in Figure 12-11. Three of them put together are shown in Figure 12-12 and a larger set in Figure 12-13. In the last Figure, all the blue tiles have been mirrored (flipped over). We suggest creating physical tiles from the materials in Christian Lawson-Perfect's repository and playing with them. Building out a pattern is harder than it looks, and addictive. Try to look at the vertices and reason about what is happening at each one. A convex vertex should be thought of as an angle greater than 180°. For the vertices where a tab pokes into a concave area on the other tile, the angle of the convex "tab" plus that of the hole around it will add up to 360°.

FIGURE 12-12

Three hats fitted together

FIGURE 12-13

Pattern formed with many hats

Finally, although it is not strictly speaking a tiling, you might want to play with a 2,000-year-old tile puzzle attributed to our friend Archimedes, the ostomachion. You can cut out its 14 pieces from one of the versions online (for example, in the Wikipedia entry "Ostomachion") and explore the various ways the triangles and quadrilaterals can be put together. People have also created versions of it on the model-download sites **printables.com** (https://www.printables.com/) (as "Archimedes puzzle") and **thingiverse.com** (https://www.thingiverse.com/).

Similar 7-piece puzzles called *tangrams* have been around in China for millennia, and have been popular in the West as entertainments for a long time too. They (and the ostomachion) belong to a broader class called *dissection puzzles,* if you want to explore even more. Playing with one (or, even better, trying to create one) is a great way to build your intuition about relationships among different simple shapes.

Robot Control

To this point in the chapter, we have looked at ways basic trigonometry relationships allow us to measure distances and angles. Being able to do that allows us to know how big an object is, or how far away it is. Or we can navigate somewhere using relative positions of the Sun and stars, the horizon, or multiple spacecraft. But we can go beyond observing positions of distant objects. We can also use the same math to control the motion of robots.

Controlling how a robot arm needs to move to get it to some desired final position is called *inverse kinematics.* It involves starting with a target arm position, then working out how all the arm's joints need to move to get to that position. If we have a simple robot arm that can only move by turning three independent servo-controlled joints, there is a limited set of ways to get to any desired endpoint. Let's see how to build the robot hardware, and then how to use trigonometry to program it.

As a side note, more complex robots might allow an arm to get to a given position many different ways. Then, math from the discipline of *linear algebra,* and possibly others, may be necessary to figure out how to move the arm in a way that is optimum for power, safety, or other constraints.

Our simplified few-inch-tall robot arm, though, is designed to be easier to deal with so we can build it with hobby parts and math we already know. Like a human arm, it moves by rotating joints. Unlike a human arm, however, the robot we are building uses a total of three servos (Figure 12-14) each of which will allow turning in only one axis. Building this robot requires access to a 3D printer, and some knowledge of setting up electronics using an Arduino microprocessor. One could also use a Micro:bit or other processor, with different code and a different power distribution scheme.

FIGURE 12-14

Completed robot arm

Creating this robot arm is an intermediate electronics project. It does not require any soldering, but does involve things like stripping power wires and attaching them to screw terminals. If you would prefer not to build the robot yourself and just want to read and follow along to see how we use trigonometry, skip straight to the "Analyzing the Arm Motion" section.

Be sure you know about safety procedures for working with electronics. Wash your hands after handling boards and do not eat and assemble at the same time, in case the boards have some lead or other substances you do not want to eat present.

If you would like to have enough background to do this yourself, but you have no robot experience, you should find an Arduino robotics book. A few good ones are *Arduino Robotics* by John-David Warren, Josh Adams, and Harald Molle (2011, Apress) and *Arduino Workshop* by John Boxall (2013, No Starch Press). Although both these books have been out for a while, the basics have not really changed much since then.

Building the Robot Arm

The Arduino microprocessor has been an open source standard for a long time, ideal for running small projects (and some big ones). We will use it to build our robot. If you have not worked with it before, check out the getting started documentation at **docs.arduino.cc** (https://docs.arduino.cc/). The software library to run servos from an Arduino, and directions on how to install them on your computer, can be found in their **servo reference materials** (https://arduino.cc/reference/en/libraries/servo/) online. As with all electronics builds, use caution and read basic tutorials carefully before proceeding.

Servos (Figure 12-15) are motors that turn a shaft to a commanded angle. That is, giving a command to a servo will make a shaft turn until the piece on top of the shaft (called a servo *horn*) points in the commanded direction. You can learn a lot more about servos, and their limitations, at retailer **Sparkfun** (https://learn.sparkfun.com/tutorials/hobby-servo-tutorial)'s tutorial or by searching "hobby servos" online. The servos we are using here can only move through half a circle (180°, or π radians). Three of them will be used to turn our robot arm in three dimensions.

FIGURE 12-15

A servo motor

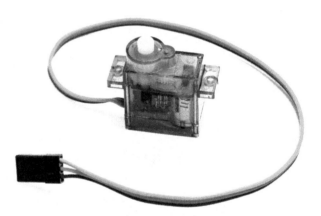

To create the robot, first 3D print the four pieces created by `simple_arm_sg90.scad`. This file has several parameters, but it would be best to leave them at their default values for an initial build (Figure 12-16). The OpenSCAD model has the parameters `bicep_l` and `forearm_l` for the lengths of the bicep and forearm segments. If you change these, remember to also change them in the Arduino code.

Do not scale the models, since they are designed to exactly fit around servos and their horns. We recommend printing one piece at a time, starting with the "bicep" piece, and making sure it fits the servo body and the servo horn before proceeding. If you find that the fits are too tight or too loose with your particular printer, you can use the "Horizontal Expansion" (Cura) or "XY Size Compensation" (PrusaSlicer) settings to adjust the fit. These settings work by moving the print's walls in or out, without changing the distance between features.

FIGURE 12-16

3D printed robot parts (Bottom, middle and top)

Next, gather up the electronics and tools you will need: an Arduino, a servo controller board compatible with it (Figure 12-17), three servos, jumper wires, the battery case, screwdrivers and, if needed, wire strippers (if the battery pack does not already have wire suitable for a terminal connector).

FIGURE 12-17

The servo controller board

The arm should be assembled from the ground up, starting with the base piece. Trying to assemble it out of order will make some screws difficult to access.

First take the base 3D printed part (the large X-shaped piece) and one of your servos. Feed the cable through from the top, then press the body of the servo into the hole, making sure that the shaft points up from the center of the X. The servo cable should come out of the cutout at the bottom of the piece. Secure the servo with the two pointed screws that came with it (Figure 12-18).

FIGURE 12-18

Creating the base

Next, take the shoulder piece and press one of the cross-shaped servo horns into the base, shaft-side first, then press it down onto the servo's *spline* (the output shaft of the servo, which has a toothed bit that fits into the horn). Before securing the horn with a screw, use it to turn the servo back and forth to find its limits. Try to turn it to its center position, between the two extremes, then lift the piece back off and re-insert it facing the direction you want to be the arm's "neutral" position. This does not need to be exact, but unlike the other joints, you will not be able to adjust this angle easily after the following steps.

Insert the non-pointed screw into the hole in the middle of the servo horn, and tighten it onto the shaft (Figure 12-19). Check that the pressure is not adding significant resistance to turning the servo. If it is, back the screw out about 1/8 turn, then pull up on the shoulder piece. Repeat this until the resistance feels about the same as before you inserted the screw.

FIGURE 12-19

*Screwing down the
first servo horn*

Next, place the second servo body sideways on top of the first servo horn (Figure 12-20). Be sure that this servo's shaft is oriented toward the center of the piece, directly above the first servo's shaft. Its wires should face toward the center of the printed piece. Secure it with two more pointed screws.

FIGURE 12-20

*Adding the second
servo*

Take the second cross-shaped servo horn and insert it onto the servo spline. Rotate it to find the extremes, then try to turn it to its center point. We can adjust the exact position later. Remove the horn from the servo, and insert it into the printed "bicep" piece. Place it back onto the servo pointing straight up, and secure it with the non-pointed screw (Figure 12-21).

FIGURE 12-21

Adding the second servo horn

Take your third servo and feed its wire through the hole in the end of the bicep piece from the side where the shoulder servo is attached. Insert the servo body through the hole, with the servo shaft and wires facing upward, away from the shoulder joint (Figure 12-22). Secure it with two pointed screws.

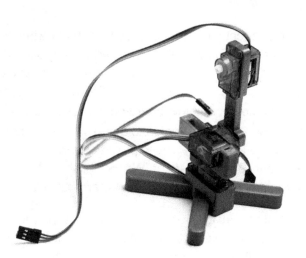

FIGURE 12-22

Adding the third servo

Repeat the centering process for the final servo, then place the horn into the forearm piece, and secure it to the elbow servo pointing straight up. Secure it with the final non-pointed screw. Turn the shoulder and elbow joints by hand, loosening the screw in the servo horn if necessary to allow them to turn freely (Figure 12-23).

FIGURE 12-23

The completed robot arm

Wiring the Arm

Now we need to connect the wires. Servos have three wires that are intended to carry a signal (orange), power (red), and ground (black). You can drive a servo directly from most microcontrollers, but you get better control with a dedicated PWM driver board. "PWM" stands for *pulse width modulation*, which means that this board generates precisely timed pulses that are used to specify angles to a servo. This board interfaces between the Arduino and the servos to provide more precise pulses than the Arduino's built-in PWM generator. We call this board the "servo driver board" for the most part, since that is how we are using it.

This board also interfaces to the (separate) power supply for the servos, and gives us convenient places to plug them in. Servos can draw more current than the regulators on some microcontroller boards are designed to handle, and some USB connections do not provide enough current for them. If this happened while the servos were powered from the same source as the microcontroller, it could cause the controller to brown out.

The servo driver board distributes power to the servos and generates control pulses as instructed by the microcontroller. Note that most microcontroller boards can generate servo control pulses directly, but they are less precise than the 12-bit PWM generator on the driver board. Distributing power to the servos is also trickier without this board.

Connect the wires from the three servos to the corresponding colored pins on the servo controller board as shown in Figure 12-24. Attach the wires from one servo to one row of pins on the controller board. Red wires go to red (power) pins, black to black (ground) pins, and orange to yellow (signal) pins. The base servo should connect to the row labeled 0, the shoulder to row 1, and the elbow to row 2 (Figure 12-25).

FIGURE 12-24

Closeup of the connections from servos to servo controller board.

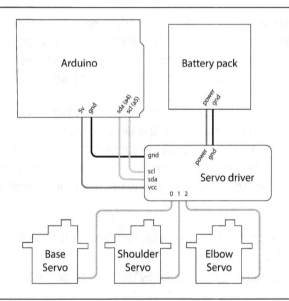

FIGURE 12-25

Wiring diagram for the robot arm

Next, we need to provide power for the servos through the servo driver board. The servos need about 5 volts, which is the voltage that USB provides. However, it is inadvisable to attempt to drive servos from your computer's USB port, since the servos may attempt to draw too much current.

We use an AA battery pack to power the servos. These come with bare wires at the ends to connect to the screw terminals on our servo driver board, as shown in Figure 12-25. Three fully charged alkaline batteries in series would give us almost exactly 5 volts, but we are going to be using rechargeable AA batteries. Nickel-metal hydride rechargeable batteries have a lower maximum voltage, so we use four of them. Do not put the batteries in the battery pack until you are finished with all assembly though.

Now we need to connect the servo control board to the Arduino. The servo controller we have chosen to use just needs to get power, ground, and a two-wire data signal for all three servos, since this control board uses the *i2c protocol* (officially i^2c, pronounced "i-squared-c", really iiC, for *inter-integrated circuit protocol*). The two signal wires are called "scl" (serial clock) and "sda" (serial data), and are often colored yellow and green, respectively. The protocol is intended for devices at close range (like those within one device) to talk to each other. There is a nice tutorial on **Sparkfun's site** (https://learn.sparkfun.com/tutorials/i2c).

The chip on the servo controller board receives these commands for the servos and generates the appropriate PWM signals for the three servos. This servo driver board also distributes power from its power input (or V+, the battery pack in this case) connection to its 16 servo outputs.

Now unplug the Arduino board (if you had it plugged into your computer) to get ready for connecting it up to the servo board. Then, using four jumper wires (thin wires with a connector on each end — you will need ones that are female on one end and male on the other), connect the servo controller board to the Arduino as follows (and as shown in Figure 12-25).

- Connect any of the Arduino pins labeled "gnd" to the "gnd" pin on the servo driver board.
- Connect the Arduino pin labeled "5v" to the "vcc" pin on the servo driver board.
- Connect the Arduino pin labeled "a4" to the "sda" pin on the servo driver board.
- Connect the Arduino pin labeled "a5" to the "scl" pin on the servo driver board.

The servo driver board also has pins labeled "OE" and "V+", which we will leave disconnected. Figure 12-26 shows the final arrangement on the servo controller board end, and Figure 12-27 shows the Arduino connections. Finally, connect the Arduino to a computer via USB. Now we need to load code onto the Arduino to make it work.

FIGURE 12-26

Closeup of the servo controller board end of the jumper wires

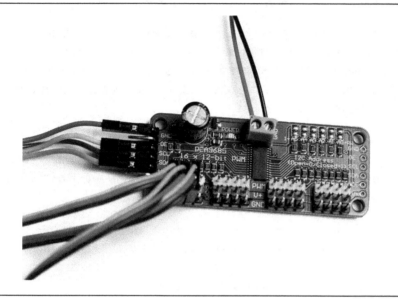

FIGURE 12-27

Close up of the Arduino end of the jumper wires

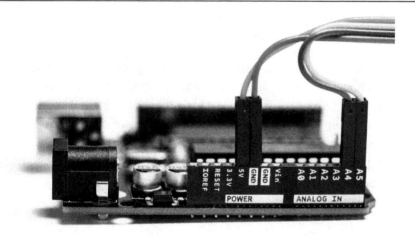

Programming the Arm

Now, our robot arm needs to be told what to do. We used the Arduino Integrated Development Environment (IDE) to write the control code for this example. You can find downloads to set up this environment, and instructions, at **docs.arduino.cc/software/ide-v2** (https://

docs.arduino.cc/software/ide-v2). If you prefer to use a browser-based approach, check out the options at **arduino.cc/en/software** (https://www.arduino.cc/en/software).

You will also need to install servo libraries (`servo.h`), as described at **arduino.cc/reference/en/libraries/servo** (https://www.arduino.cc/reference/en/libraries/servo/). How this works, and how you will tell your computer which Arduino board is attached, will vary depending on which of these systems you elect to use.

If, like us, you use an i2c motor controller between the servos and the Arduino, you will need to include the `wire.h` library in your Arduino code. This includes (among other things) the software for the Arduino to talk to the motors with this protocol.

Finally, the Arduino libraries do their math in radians. However, the standard servo library expects degrees, so we needed to add strategic conversions. The Arduino code we will use to experiment with our configuration is also in this book's Github repository (see Chapter 2) as the file `servo_arm.ino` , and is listed in an upcoming sidebar.

Analyzing the Arm Motion

After all that, we are finally ready for some trig. We want to input the desired positions of the tip of the robot arm in Cartesian coordinates. However, since servos can only rotate and not translate, it is easier if we convert to a spherical coordinate system (Chapter 4) and define motion in (r, θ, φ) instead. The axes of rotation of the two base servos intersect at the origin of that coordinate system. The (r, θ) plane is centered over the base, at the height of the shoulder servo. Figure 12-28 shows how a spherical coordinate grid lines up with the arm.

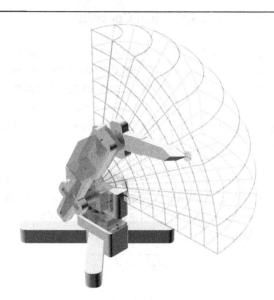

FIGURE 12-28

The robot and a spherical coordinate grid

The base servo will rotate in the θ direction from 0 to π (or 0 to 180°). It is limited to half a circle because these servos only move that much. That means that other servos need to turn to get to other points in the hemisphere.

The middle ("shoulder") servo controls motion in the φ dimension, again from 0° to 180°, or 0 to π radians). We have defined $\varphi = 0°$ to be straight up (i.e., the plane below the base would be 90°).

The uppermost ("elbow") servo allows for adjusting the radius variable, r, and flips over to get to the other values of θ that the base servo cannot reach.

In all cases, though, because servo code expects angles between 0° and 180°, we need to convert to angles in that range.

If we want to have the point of the robot's arm touch some point in space, in general, we will need to turn all the servos so that the point touches the desired spot. If we are specifying a point (or curve) in Cartesian coordinates, first we will need to change those to spherical coordinates, to correspond to the three servo motor directions.

Converting Cartesian to Spherical Coordinates

Converting from Cartesian to spherical coordinates in general is algebraically a little ugly. It also requires deciding on several arbitrary coordinate conventions. If we imagine Earth in spherical coordinates, we have to decide whether the north pole or the equator is $\varphi = 0°$. In this project, we assume that $\varphi = 0°$ is the "north pole". The actual Earth has the equator as $0°$.

When we program our robot arm, we are only converting from a desired position (in Cartesian coordinates) to commanded angle changes (in spherical coordinates) so we do not have to worry about converting back to Cartesian coordinates. The conversions are

$$r = \sqrt{x^2 + y^2 + z^2}$$

$$\theta = \tan^{-1}\left(\frac{y}{x}\right)$$

$$\varphi = \cos^{-1}\left(\frac{z}{r}\right)$$

The "radius" coordinate is something we think of as a straight line from the center of our coordinate system to the point we are describing. We do have to deal with some complications, though, because of physical limitations of the servos, which we will get to shortly. We also have to be careful of how we calculate arctangent, which is a little tricky since we need to keep track of the sign of the result.

The atan2 Function

Our θ angle, as we saw in Chapter 5's discussion of polar and spherical coordinates, can take on any value from $0°$ to $360°$. However, the basic arctangent function to find θ returns an angle between -90° and 90° (the principal value we learned about in Chapter 5). This is handled by using the `atan2(y, x)` math function instead of ordinary single-argument arctangent, `atan(x / y)`.

The `atan2(y, x)` function takes in a pair of coordinates (note the order: y first) and, by preserving the sign of both arguments, can correctly return angles outside the range of basic arctangent. It will return the angle between the x axis and a line from the origin of the Cartesian coordinate system to the point (x, y). The Arduino code will use `atan2(y, x)` to compute θ. This function has been around for a long time and is supported in most major programming languages, should it come up in other problems you need to solve. For its history and more on the various special cases it handles, check out the "atan2" Wikipedia article or documentation for any system you might be using later. This function will return an answer in radians, not in degrees.

Walkthrough of the Arduino Code

The Arduino sketch (as these programs are called) is listed in full in the sidebar. Arduino programming is done in the C programming language. Let's briefly walk through what it does, based on the background we have seen in the last few sections about how this arm works.

The sketch first initializes and defines the various pieces of hardware and tells the rest of the code what is attached to what. Then, various functions are defined. Notably, the function `cartesian` calculates servo angles given a position in Cartesian coordinates, and moves the arm accordingly. Some adjusting of the output is required, as we noted in the previous section, so that the servo angles will fall between 0° and 180°.

Awkwardly, some library functions we are using (notably atan2) return values in radians, but others (like the input to the servos) need to be in degrees. In the code, we converted radians to degrees right away and have for the most part calculated in degrees.

Once we convert from Cartesian to spherical coordinates, we need to calculate the angles for three servos, which we call `base`, `shoulder`, and `elbow`. The `base` servo angle is easy to calculate. It is just θ, though we need some additional logic to handle angles beyond 180°.

The `elbow` angle is also straightforward since we only need to know the radius and the lengths of the `bicep` and `forearm` segments. These three lengths form a triangle. To calculate one angle of a triangle from three known sides, we use the law of cosines (described in Chapter 6). As shown in Figure 12-29, we know the lengths of the two plastic arms that make up the sides of the triangle, as well as knowing r from the given desired final position of the arm. This lets us compute the `elbow` angle.

FIGURE 12-29

The triangle defining φ

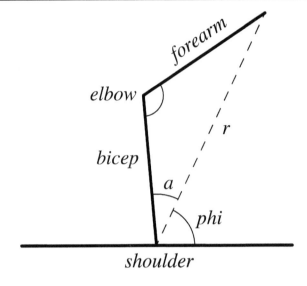

Calculating the `shoulder` angle is a bit more complicated. For this, we need to use φ. With the `elbow` joint fully extended, the `shoulder` angle equals φ. However, when the arm is bent, we need to add the angle labeled *a* in Figure 12-29. In other words, a + φ is the `shoulder` angle. But what is the value of angle *a*? Since we have three sides of the triangle, we could just use the law of cosines again. It is a bit simpler though to use the law of sines (also in Chapter 6), now that we know one of the angles. This is how the sketch computes this angle.

All Arduino programs, once functions and variables are set up, have a program consisting of two parts: `setup()` and `loop()`. The code in `setup()`, as the name implies, is run only when the program starts. The `loop()` function is run over and over until the Arduino is turned off. In this sketch, the `loop()` function draws a square in space over and over. You can play with altering that to see what other shapes the tip of the arm is able to draw.

The `BASE_OFFSET`, `SHOULDER_OFFSET`, and `ELBOW_OFFSET` variables allow you to fine-tune the angles of the servos. If one of them is more than a few degrees off, though, you need to unscrew the servo horn and remount it, because the servo's range of motion is limited. Our example uses the ServoEasing library, which you can download through the Arduino library manager. The `Servo hardware abstraction` section of our sketch can be modified to use a different servo library if you want to adapt the code for another controller.

Arduino Sketch for Robot Arm

```cpp
#include <Arduino.h>

//Calibrated angle offsets for each servo:
#define BASE_OFFSET 0.0f
#define SHOULDER_OFFSET 1.0f
#define ELBOW_OFFSET 1.0f

//Lengths:
#define BICEP 50
#define FOREARM 50
#define SQUARESIZE 65

//Servo hardware abstraction:
#define BASE_PIN 0
#define SHOULDER_PIN 1
#define ELBOW_PIN 2
#define USE_PCA9685_SERVO_EXPANDER //used by ServoEasing.hpp
#include "ServoEasing.hpp"
class aServo {
  ServoEasing thisServo;
  public:
    aServo() {
      thisServo = ServoEasing(PCA9685_DEFAULT_ADDRESS);
    }
    void go(float angle) {
      thisServo.setEaseTo(angle);
    }
    void setup(int pin, float offset) {
      thisServo.setEasingType(EASE_CUBIC_IN_OUT);
      thisServo.setSpeed(500);
      thisServo.attach(pin);
      go(90);
    }
};

void syncServos() {
  //Add a delay here for libraries without a wait function
  synchronizeAllServosStartAndWaitForAllServosToStop();
}

//End servo hardware abstraction
```

```
float r2d(float r) {return r * 180 / PI;}
float d2r(float d) {return d * PI / 180;}

aServo base;
aServo shoulder;
aServo elbow;

void cartesian(float x, float y, float z) {
  float r = sqrt(pow(x, 2) + pow(y, 2) + pow(z, 2));
  float theta = r2d(atan2(y, x));
  float phi = r2d(acos(z / r));

  Serial.print("r = ");
  Serial.println(r);
  Serial.print("theta = ");
  Serial.println(theta);
  Serial.print("phi = ");
  Serial.println(phi);

  //law of cosines, C = acos((a^2 + b^2 - c^2) / 2ab)
  float elbow_a = acos((pow(BICEP, 2) + pow(FOREARM, 2) - pow(r, 2)) /
    (2 * BICEP * FOREARM));
  //law of sines, A = asin(a sin(B) / b)
  float shoulder_a_adjust = r2d(asin(FOREARM * sin(elbow_a) / r));
  elbow_a = r2d(elbow_a) - 180;

  Serial.print("elbow = ");
  Serial.println(elbow_a);

  if(theta < 0) {
    theta += 180;
    phi *= -1;
    elbow_a *= -1;
    shoulder_a_adjust *= -1;
  }

  base.go(theta);
  shoulder.go(90 - phi + shoulder_a_adjust);
  elbow.go(90 - elbow_a);
  syncServos();
}

void setup() {
  Serial.begin(9600);
  base.setup(BASE_PIN, BASE_OFFSET);
  shoulder.setup(SHOULDER_PIN, SHOULDER_OFFSET);
```

```
  elbow.setup(ELBOW_PIN, ELBOW_OFFSET);
  delay(500);
}

void loop() {
  for(int i = -SQUARESIZE; i <= SQUARESIZE; i += 2) {
    cartesian(i, -SQUARESIZE, -30);
  }
  for(int i = -SQUARESIZE; i <= SQUARESIZE; i += 2) {
    cartesian(SQUARESIZE, i, -30);
  }
  for(int i = -SQUARESIZE; i <= SQUARESIZE; i += 2) {
    cartesian(-i, SQUARESIZE, -30);
  }
  for(int i = -SQUARESIZE; i <= SQUARESIZE; i += 2) {
    cartesian(-SQUARESIZE, -i, -30);
  }
}
```

Running the Robot

To finish this project, check all the connections one more time, and then put batteries into the battery pack. Finally, plug your Arduino into a laptop running the Arduino development environment. Download the robot arm sketch to the Arduino, and run the sketch. Note that the robot arm should move right away and start drawing a square in the air, so be ready for that. You can experiment with changing the sketch to draw different patterns.

Building on this Project

Now, what can you do next? A robot arm that just points to a spot is not the most interesting application. Once you know how to do that, though, you might add a gripper or other tool to the end of the arm. Or, with a little modification, you can flip the arm over and turn it into a leg, to make a robot that walks like the one in Figure 12-30.

FIGURE 12-30

A walking robot.

Looking Forward to Calculus

In this chapter, we focused on tying together ideas from previous chapters to enable us to try a few more-ambitious projects. But where do we go from here? Trigonometry and analytic geometry underlie most of engineering mathematics, starting with the basics of calculus.

Sometimes topics in this book are gathered up with other topics to create a "precalculus" course. Too often, those courses are lots of drill in algebra techniques which, although they can be useful in later math, are out of context. That makes them hard to understand without some intuition about why they are useful. We have tried to walk the line of having just enough algebra so that our work is useful as a supplement to a traditional algebra-first course.

The conic section chapters are a bridge between the geometry-focused basics of trigonometry and the analytic geometry analysis of how the properties of a curve are reflected in the equations for and graphs of it. Should you move on to our *Make: Calculus* book, we then show how we can tie these curves (and equations) to real applications. Ultimately this lets us predict how things move and change in the real world. Parabolas come up in many applications, and in calculus we learn where some features just asserted here come from (like the coordinates of parabola's vertex).

Trigonometric functions like sine and cosine are everywhere in calculus (and engineering and scientific applications). In calculus we learn that many periodic real systems can be modeled with one or many sinusoids, and some features of them make them very handy to work with in calculations. In calculus you will learn where the equations for swinging pendulums and other periodic systems come from. If you go on to electricity and magnetism and its associated math, you will see a lot more about waves and how they propagate too.

This book was intended to connect our *Make: Geometry* and *Make: Calculus* books. We hope we have given you the confidence to march right across that bridge, and keep going!

Chapter Key Points

In this chapter, we explored several different tiling projects to explore how polyhedra fit together. Then we embarked on a relatively complex robot arm project, requiring pieces from several earlier chapters. This exercise explored inverse kinematics, the process of figuring out how to use a robot's motors to move its parts to a desired point in space. Finally, we summarized how ideas in this book set us up for learning calculus next, and surveyed useful resources.

Terminology and Symbols

Here are terms we used in this chapter which you might want to look up for more information:

- Arduino
- aperiodic tiling
- atan2(x) function
- dissection puzzle
- einstein
- i2c
- inverse kinematics
- monotile
- ostomachion
- pulse width modulation (PWM)
- servo
- tangram
- tessellations and tilings

Resources

Sources for this chapter, as well as some we used in many places in this book, are collected here. There are many excellent mathematics websites and resources, but these are ones we have found particularly helpful.

Sources for this Chapter

Arduino resources can be found at **docs.arduino.cc** (https://docs.arduino.cc/). Searching on that site is likely to bring up the other resources you need. There are also other good resources at sparkfun.com and adafruit.com, although those may be aimed more at their implementations of the Arduino standards. More detailed web references are listed in context in the chapter.

A good general introduction to tiling can be found in this article by Natalie Wolchover in *Quanta* magazine (July 11, 2017): "Pentagon Tiling Proof Solves Century-Old Math Problem", **retrieved July 3, 2023** (https://quantamagazine.org/pentagon-tiling-proof-solves-century-old-math-problem-20170711/).

There has been a lot of press just as we were writing this book about the einstein monotile. Besides the preprint that started it all (web link in the chapter narrative), good general overviews can be found at Will Sullivan's March 29, 2023, piece in *Smithsonian* magazine, "At Long Last, Mathematicians Have Found a Shape With a Pattern That Never Repeats, **retrieved July 3, 2023** (https://smithsonianmag.com/smart-news/at-long-last-mathematicians-have-found-a-shape-with-a-pattern-that-never-repeats-180981899/).

Websites and Search Terms

Wikipedia (https://wikipedia.org) is a good resource for reference material like tables of integrals, derivatives, and trigonometric identities. Web searches for phrases like "integral table" will likely return many hits, which of course need to be reviewed for accuracy. We have found Wikipedia to be generally reliable this way, if occasionally idiosyncratic with (unexplained) notation which can make things challenging for a beginner. Since pages in Wikipedia change all the time, your best bet is to search a bit. If you are doing something with real-world implications, we suggest investing in one of the resources we have listed in the Books section.

Greg Sanderson's online video series, "3Blue1Brown" (available on **youtube.com** (https://www.youtube.com/) or through **3blue1brown.com** (https://www.3blue1brown.com/)) has many innovative ways of viewing mathematics topics. We particularly found his discussions of the complex plane inspirational.

The Khan Academy (**khanacademy.org** (https://www.khanacademy.org/)) is a well-organized place to look for background on almost any mathematics subject.

If you want to learn about math concepts with a more physics bent, check out Georgia Tech's **Hyperphysics** (http://hyperphysics.phy-astr.gsu.edu) site.

Calculation Resources

If you want a graph of a function, type it into the **Google** (https://google.com) search box in pseudocode. Use `*` for "multiply", `/` for divide, `+` for "add", and `^` for raise to a power. So, to get a graph of $y = x^2$ for instance, you would type `y = x^2` in the search box. For

$y = 2x$, you would type `y = 2 * x`. We found that `y = 2x` also seems to work, but to be safe we always use the `*`.

You can zoom in and out by scrolling and move around to see other parts of the graph by dragging a mouse. More sophisticated queries work as well, but we have seen (rare) errors in these results. If you are doing something important be sure to check search engine calculation results by another means.

The math site **Desmos** (https://desmos.com) also has many free resources and downloadable apps for graphing and calculating. More generally, you can search "integral calculator" or "derivative calculator" or just "calculator" to find a plethora of free resources.

The **Wolfram Alpha** (https://wolframalpha.com) site has a mix of free and paid symbolic calculation tools, including the ability to type in an integral or derivative and get a graphical or symbolic answer. Wolfram's **Mathematica** (https://wolfram.com/mathematica/) is a pricier and more capable option. **Wolfram Mathworld** (https://mathworld.wolfram.com) has many definitions of mathematical concepts.

Finally, if you want to be able to do analysis without necessarily having an internet connection, in addition to those mentioned so far there are many paid and free apps that work similarly. For example, both of us use MacOS computers, which have for some time come with an app called "Grapher" that has limited integration and derivative graphical capabilities. Note that Grapher gives incorrect answers or will not work if spaces are not used when it expects spaces. Experiment with a known equation to check the conventions it is using.

Most of the graphs in this book were created with Grapher and then exported to Adobe Illustrator to make them better for publication. Microsoft Excel (or other spreadsheet tools) can also be very useful if you just want to play around with graphing data, or generate data to graph. You can of course tweak some of the OpenSCAD models in this book to do 3D visualizations. It is also possible to do animation in OpenSCAD. (Search online for "animate with OpenSCAD" to find tutorial videos.)

Books

If you found yourself a little hazy on some of the underlying geometry of our approach to trigonometry, you might benefit from our 2021 book,

Make: Geometry, also from Make Community LLC. Likewise, if you need a little help with 3D printing per se, you might find our *Mastering 3D Printing, 2nd Edition* (New York: Apress, 2020) useful. Of course, if you want to keep going, you can explore our 2022 *Make: Calculus* book.

We found the following resources broadly useful in our research. References that are more specific to a particular chapter are also found at the end of those chapters. We annotated why we have highlighted each book.

Boaler, Jo, *Mathematical Mindsets.* San Francisco: Jossey-Bass/Wiley, 2016. Discusses how to teach math in general in a more interactive way.

If you like playing with mathematics, you might enjoy Martin Gardner's works. Our book technical reviewer credits Gardner's "Mathematical Games" column in *Scientific American* magazine for showing him that math was both useful and fun. *Mathematical Carnival* (Vintage Press, 1977) might be a good book to peruse first. He also suggests the works of Eric Temple Bell, as well as James R. Newman's four-volume *The World of Mathematics,* available in various editions new and used. The **Mathematical Association of America** (https://www.maa.org) and the **American Mathematical Society** (https://www.ams.org) also have catalogs of many interesting books to take you further.

Lockhart, Paul. *A Mathematician's Lament.* New York: Bellevue Literary Press. 2009. Lockhart analyzes what is wrong with common methods of teaching math and how to encourage intuition over problem-solving. If you teach math, you will appreciate its pragmatic points of view. His other books *Measurement,* Cambridge MA: The Belknap Press of Harvard University Press, 2012; and *Arithmetic,* Cambridge MA: The Belknap Press of Harvard University Press, 2017, apply some of his ideas.

Mahajan, Sanjoy. *Street-Fighting Mathematics: The Art of Guessing and Opportunistic Problem Solving.* Cambridge: The MIT Press,2010. This is a great introduction to developing intuition to solve problems rather than leaping immediately into calculation, using dimensional analysis and other techniques we have touched on in this book. You might also try searching for Mahajan's 2011 Caltech TEDx talk where he demonstrates some of the techniques.

Merzbach, Uta C., and Boyer, Carl B. *A History of Mathematics, Third Edition*. Hoboken, NJ: John Wiley and Sons, 2011. This comprehensive history of mathematics does not teach the principles as it goes, so might be heavy going for readers of this book. However, if you would like to dip around now that you have some background, you might enjoy going into a bit more depth.

van Brummelen, Glen. T*he Doctrine of Triangles: A History of Modern Trigonometry*. Princeton, NJ: Princeton University Press, 2021. This history of trigonometry presumes the reader is a sophisticated mathematician already, but was a good reference for us to clarify the history of some of the innovations we describe in this book.

Zwillinger, Dan. *CRC Standard Mathematical Tables and Formulas, 33rd Edition*. Boca Raton: CRC Press/Taylor and Francis, 2018. If you want to look up any math formula and have a validated source for it, this is the book for you. Although it is very hefty, if you are doing real-world problems where the answers have to be right it is an indispensable resource. The 2018 edition has sections on numerical methods and a section on the classic equations from many disciplines.

Topics Covered

Although we have not designed this to be a textbook covering all of any particular course, we hope it will be useful as a supplemental book and resource for those in a variety of learning environments. As such, we have not tied ourselves to any particular set of standards, but rather to topics and concepts to make it more generally applicable wherever our readers may be.

If you would like to reference a broad set of standards on what topics to teach and when, you might search online for "Common Core Standards" which we retrieved from **learning.ccsso.org/common-core-state-standards-initiative** (https://learning.ccsso.org/common-core-state-standards-initiative), or try searching online with "(your state or other jurisdiction) math standards" for more local inputs. As of this writing there was some discussion about standards in many jurisdictions, but we are sure that when that settles down you will be able to map these topics to whatever the future brings in that regard.

The following lists the key topics in each chapter:

- Chapter 1. Trigonometry and Analytic Geometry
    - Definitions of trigonometry and analytic geometry
    - What algebra and geometry you should know already
- Chapter 2. OpenSCAD and the 3D Printed Models
    - Downloading OpenSCAD
    - Downloading this book's models
    - OpenSCAD capabilities
    - How to edit a pre-existing file in OpenSCAD

- Chapter 3. Triangles and Trigonometry
  - Converting degrees to radians and vice versa
  - Proving that the angle of a triangle add to 180°
  - Using a protractor
  - Conventions for labeling triangle sides and angles
  - Acute, right, and obtuse triangles
  - Scalene, isosceles, and equilateral triangles
  - Understanding congruence and similarity of triangles
  - Applying similarity to the problem of measuring a large object
  - The Pythogorean Theorem
  - Square root notation
  - The Spiral of Theodorus as an analog square root calculator
  - Definition of basic trig functions: sine, cosine, tangent, secant, cosecant, cotangent, and their inverses
- Chapter 4. Coordinate Systems and Analytic Geometry
  - Cartesian and polar coordinates
  - Coordinate of a point
  - Slope and intercept of a line
  - 3D coordinates: Cartesian, cylindrical, spherical
  - Plotting a curve or surface in 3D
- Chapter 5. The Unit Circle
  - Trig functions for angles above 90°
  - Principal value of an angle
  - The unit circle and continuous trig functions
  - Phase, frequency and amplitude of a trig function
  - Angular frequency
  - Period
  - Graphing trig functions in Cartesian and polar coordinates

- Chapter 6. Trig Identities to Logarithms
  - Law of Sines and Law of Cosines
  - 45-45-90 and 30-60-90 triangles
  - Trig cofunction relationships (cosecant, cosine, cotangent)
  - Complementary angles
  - Trig identity of squares of sine and cosine
  - Sum of angles and double-angle formulas
  - Prosthaphaeresis and transforming multiplication calculations into addition
  - Logarithms base 10
  - How a slide rule works
  - Calculating cube and square roots with logs
  - Operations with logs (multiplication, division, raising to a positive or negative power)
  - Estimation using log plots
- Chapter 7. Navigation
  - Making and using an inclinometer
  - Measuring angles and distances with trig
  - Finding latitude by sighting Polaris
  - Trig applications in navigation using astrolabes, sextants, and GPS
- Chapter 8. Making Waves
  - Nonsinusoidal periodic waves (sawtooth, square)
  - Superposition of waves
  - Constructive and destructive interference
  - Effects of wavelength and frequency on wave propagation
  - Point source model
  - Electromagnetic waves
  - Refraction and Snell's Law
  - Rotated coordinate systems
  - Lenses and refraction
  - Reflection and Stokes' Relations
  - Water and sound waves
  - Helicoid properties, as a minimal surface
  - Soap bubble demonstration of a minimal surface

Appendix A, Topics Covered

Index

Symbols

A

B

Briggs, Henry, 121, 122, 123

C

Cameron, Rich, xvii
Cartesian coordinate system, 57
  3D, 65
  analogy to New York streets, 59
  converting to polar coordinates, 69
  converting to spherical coordinates, 313
  graphing a line in, 63
  model of, 62
    (see also 3D printed models)
  slope of a line, 63
Cassegrain focus, 279
Cassegrain telescope, 279
Cassegrain, Laurent, 279
chromatic aberration, 278
clinometer (see inclinometer)
cofunction
complementary angles, 46, 114
completing the square
cone
  slant angle of, 192
  slant height of, 193
conic sections, 191, 192
  circle, 195
  circumference
    ellipse, 218
  cut angle of, summary, 199
  ellipse, 195
    area of, 216
    eccentricity, 208
    foci, 209
    semimajor axis, 209
    semiminor axis, 209
  hyperbola, 197, 259
    asymptotes, 263, 273
    circular directrix, 264
    foci, 264
    graphing, 263

    making with Play-Doh, 197
  parabola, 196
    directrix, 225, 245
    drawing, 226
    finding minimum or maximum, 255
    focus, 196, 225
    translating, 236
    vertex of, 245
coordinate system
  definition of, 57
  rotated, 169
cosecant, 115
  definition of, 46
cosine
  3D printed model, 91
  definition of, 42
  finding with hypotenuse.scad, 44
  of angles over 90 degrees, 85
cotangent, 115
  definition of, 46
curve
  definition of, 74
  translating, 214
cylindrical coordinates
  3D printed model, 71
  analogy to Hawaiian roads, 60

D

definition of, 42
degrees (of an angle)
  converting to radians, 29
  definition of, 28
Descartes, René, 2, 58
Desmos, 323
dioptra, 146
dissection puzzles, 299
d'Alembert's Theorem (see Fundamental Theorem of Algebra)

333

U

unit circle
    definition of, 83
    rolling out curve from, 94

V

van Brummelen, Glen, 132, 325
vertex
    definition of, 27
Vieta's Formulas, 241
Viète, François, 119, 241
    (see also Vieta)

W

wave, 157
    electromagnetic, 168
    sawtooth, 158
    square, 158
    triangle, 158
wavelength, 162
waves
    dissipative, 182
    longitudinal, 182
    nondissipative, 182
    sound, 182
    tranverse, 182
    water, 182
Werner, Johann, 120
whispering gallery, 220
Wikipedia, 322
Wolfram Alpha, 323

Z

Zwillinger, Dan, 325

More from the authors:

Make:
Geometry

Learn by 3D printing, coding and exploring
Joan Horvath and Rich Cameron

Make:
CALCULUS
Build models to learn,
visualize and explore

Joan Horvath and Rich Cameron

Printed in the USA
CPSIA information can be obtained
at www.ICGtesting.com
JSHW051806100624
64551JS00011B/525